The Smiling Chatbot

Konstantin Prinz

The Smiling Chatbot

Investigating Emotional Contagion in Human-to-Chatbot Service Interactions

Springer Gabler

Konstantin Prinz
Universität Koblenz-Landau
Koblenz, Germany

Universität Koblenz-Landau, Dissertation, 2022
Vollständiger Abdruck der vom Fachbereich 4: Informatik der Universität Koblenz-Landau genehmigten Dissertation mit dem Titel "The Smiling Chatbot: Investigating Emotional Contagion in Human-to-Chatbot Service Interactions".

ISBN 978-3-658-40027-9 ISBN 978-3-658-40028-6 (eBook)
https://doi.org/10.1007/978-3-658-40028-6

This Springer Gabler imprint is published by the registered company Springer Fachmedien Wiesbaden GmbH, part of Springer Nature.
The registered company address is: Abraham-Lincoln-Str. 46, 65189 Wiesbaden, Germany

Zusammenfassung

Bedeutende Fortschritte im Bereich der künstlichen Intelligenz haben der Verbreitung darauf basierender Technologien in den zurückliegenden Jahren starken Rückenwind verliehen. KI-Technologien beeinflussen vermehrt private Kontexte aber auch den Servicesektor. Dort etablieren sich Chatbots als ein beliebtes Instrument für das Handling von Serviceinteraktionen. In Teilen scheint die Verbreitung von Chatbots und der damit verbundene Fokus auf funktionale Vorteile jedoch Erkenntnissen der Serviceforschung zu widersprechen, dass Serviceinteraktionen auch deutlich getrieben sind von emotionalen Komponenten. Eine zentrale Rolle spielt dabei beispielsweise die sogenannte Emotional Contagion, also die unbewusste Übertragung von Emotionen von Mitarbeitern auf Kunden. Vor dem Hintergrund, dass diese Ansteckung mit in der Regel positiven Emotionen im weiteren Verlauf für eine bessere Evaluation der Servicetransaktion führen kann, war es das Ziel der vorliegenden Dissertation, die Effekte positiver dargestellter Emotionen eines Chatbots zu untersuchen. Dafür wurden sechs aufeinander aufbauende Studien durchgeführt. Die Ergebnisse zeigen deutlich, dass ein Chatbot in der Lage ist, durch die Darstellung positiver Emotionen, Serviceinteraktionen anzureichern, indem er für eine Übertragung positiver Emotionen sorgt, was im weiteren Verlauf ebenfalls zu einer besseren Serviceevaluation führt. Es zeigt sich, dass die affektiven Reaktionen jedoch von der Persönlichkeit der Kunden sowie dem Erscheinungsbild des Chatbots, ausgedrückt durch ein Avatar, beeinflusst werden. Außerdem deuten die Ergebnisse darauf hin, dass die Reaktionen auch während kritischer Service Recoveries auftreten.

Abstract

Significant advances in the field of artificial intelligence have given a strong tailwind to the spread of technologies based on it in recent years. AI technologies are increasingly influencing private contexts but also the service sector where more and more service encounters are handled by chatbots. In part, however, the spread of chatbots and the associated focus on their functional advantages seem to contradict extant research findings that service interactions are also clearly driven by emotional components. A central role is played here, for example, by so-called emotional contagion (i.e., the unconscious transfer of emotions from employees to customers). Against the background that this contagion with usually positive emotions can lead to a better evaluation of the service transaction in the further course, it was the goal of the present thesis to investigate the effects of positive displayed emotions of a chatbot. For this purpose, six consecutive studies were conducted. The results clearly show that the expression of positive emotions by a chatbot enriches the service interactions by transmitting positive emotions that, in the further course, do also lead to a better evaluation of the service experienced. Moreover, it is shown that these emotional reactions are dependent on the customer's personality and the chatbot's appearance, expressed through an avatar. Furthermore, the results suggest that the reactions occur also during critical service recovery interactions.

Contents

Abbreviations

AFC	automated facial coding
AI	artificial intelligence
al.	alii/aliae (and others)
ANOVA	analysis of variance
AVE	average variance extracted
CA	Cronbach's alpha
CASA	computers are social actors (paradigm)
CFA	confirmatory factor analysis
CI	confidence interval
CMV	common method variance
CPA	customer positive affect
CR	composite reliability
EDB	emotional decision behavior
e.g.	exempli gratia (for example)
EP	empathy
ES	encounter satisfaction
EX	extraversion
EXPCB	experience with chatbots
EXPLC	experience with live chats
FACS	facial action coding system
fMRI	functional magnetic resonance imaging
HTMT	heterotrait-monotrait (method)
i.e.	id est (that is)
JAS	Job-Affect-Scale
LSD	least significant difference
M	mean

MICOM	measurement invariance of composite models
n.d.	no date
NFI	need for interaction
OE	openness to experience
p.	page
PDE	positive displayed emotions
PLS	partial least squares
RS	recovery satisfaction
SD	standard deviation
SEM	structural equation modeling
SR	successful recovery
SST	self-service technologies
VIF	variance inflation factor
vs.	versus

List of Figures

List of Tables

Introduction 1

1.1 Relevance

In today's world, in both private and business life, technologies powered by artificial intelligence (AI) are becoming increasingly powerful. According to Bellman (1978, p. 3), AI is the ability of computers to perform "[…] activities that we associate with human thinking, activities such as decision making, problem solving, learning […]" by first empirically observing and later transferring human thinking and behavioral patterns. On the one hand, the progressive success occurs because more and more data are available and can be used to train these technologies (Huang & Rust, 2018; Koch, 2018). On the other hand, these technologies also generate more data, which has led to dynamic growth in recent years. In private settings, for example, AI finds its way into many households as personal assistants like Amazon's Alexa or Apple's Siri, interacting with humans as if they were humans themselves.

The power of existing AI technologies also offers potential for customers and companies, which results in an increasing number of service encounters being handled by AI (e.g., van Doorn et al., 2017). Therefore, AI is also gaining relevance in many service settings. There are various reasons for this. First, there have been major developments in terms of the possible use cases for AI. These developments rest mainly on the improving power of new technologies and new ways to support frontline employees or even replace them, thus improving value creation by including large amounts of data during the encounter (Koch, 2018). Second, companies expect efficiency gains when using AI technologies (Marinova, de Ruyter, Huang, Meuter, & Challagalla, 2017; Riikkinen, Saarijärvi, Sarlin, & Lähteenmäki, 2018). These expectations rest mainly on automation potential due to the use of AI. The instantaneous processing of data in market

K. Prinz, *The Smiling Chatbot*,
https://doi.org/10.1007/978-3-658-40028-6_1

research may possibly be optimized by AI-based analytics (Gentsch, 2018). In addition, it is possible to take over tasks that occur with a high frequency and demand only a small degree of sophistication. The takeover of these tasks by AI allows employees to focus on other, more complex tasks (Sawhney, 2016).

Even though there are many more ways to use AI in service settings, chatbots are among the most widely spread forms that companies use. This thesis follows the suggested definition by Crutzen, Peters, Portugal, Fisser, and Grolleman (2011, p. 514), who describe a chatbot as "[…] an artificially intelligent chat agent that simulates human-like conversation, for example, by allowing users to type questions […] and, in return, generating meaningful answers to those questions." The special feature of chatbots is that they enable communication, for example, service requests, that is strongly aligned with human-to-human communication (McTear, Callejas, & Griol, 2016). Their increasing relevance for the service sector becomes clear when looking at the fact that there were 34,000 chatbots active on Facebook Messenger in 2017 (Schneider, 2017), increasing to more than 300,000 one year later in 2018 (Boiteux, 2019). The main reason for their fast growth is that chatbots enable customer service to take place around the clock (Schneider, 2017; Wünderlich & Paluch, 2017). Furthermore, intelligent agents can support or even replace human employees, thus lowering labor costs (Li et al., 2017), and also reduce a variance in performance levels by delivering a consistent level of service quality (Behera, 2016; Schanke, Burtch, & Ray, 2021).

1.2 Research Gaps

With the spread of chatbots, the way service tasks are being performed is set to change tremendously (Huang & Rust, 2018). While, on the one hand, the importance of chatbots in service is increasing, on the other hand, service encounters often include important emotional components, such as emotional contagion (Hatfield, Cacioppo, & Rapson, 1992). This automatically raises several questions that cannot currently be answered by the existing literature, and that consequently represent the research gaps of the present thesis.

Research gap 1: Can chatbots trigger emotional contagion?

First, chatbots as conversational agents without any physical representation have so far gone unnoticed regarding their ability to trigger emotional contagion. Only a few research papers have investigated emotional contagion triggered by conversational agents represented by an avatar (e.g., Tsai, Bowring, Marsella, Wood, & Tambe, 2012) or robots (e.g., Tielman, Neerincx, Meyer, & Looije, 2014). The basis of these research approaches is the so-called computers are social actors (CASA) paradigm (Reeves & Nass, 1996). It refers to humans' unconscious interaction (Katagiri, Nass, & Takeuchi, 2001) with computers as if they were human beings and social entities. While all the named studies were able to confirm the occurrence of emotional contagion, they were all limited by the fact that they relied on either an embodied conversational agent or a robot with a physical appearance. In comparison to robots, chatbots, as so-called text-based conversational agents, are usually characterized by their missing physical representation and the fact that they do not perform physical service tasks because the core aspect of the performed service is the conversation (Jain, Kumar, Kota, & Patel, 2018). This physical appearance is typically considered a central determinant in causing social reactions toward artificial entities (Duffy, 2003). The key challenge in this regard is that emotions in text-based communication are generally considered difficult to communicate (Walther & D'Addario, 2001). In recent years, emojis (e.g., 😀 or 😊), colored pictograms (Ganster, Eimler, & Krämer, 2012), in particular, have emerged as possible substitutes, which today offer a seemingly endless list of display options. Thus, the question of whether chatbots can trigger emotional contagion, taking into account that only text-based communication is available to them, remains unanswered. This is problematic in that it means that important findings are missing, regarding what features of a chatbot are responsible for triggering social responses in customers. With the progressive spread of chatbots, the question then arises as to whether, from the customer's point of view, social interaction partners are lost in service encounters or if there is a way for chatbots to adequately substitute these emotional components as well. It is therefore important to investigate the influence of human-like behavior, in the form of displayed emotions by a chatbot, on customers.

Research gap 2: Are emotional components equally as important in chatbot-handled service encounters as they are in human-to-human interactions?
Second, many studies have highlighted the importance of emotions for marketing (e.g., Grandey, Fisk, Mattila, Jansen, & Sideman, 2005; Kenning, Plassmann, Deppe, Kugel, & Schwindt, 2005; McClure et al., 2004; Pugh, 2001). In the course of research on emotional contagion (Hatfield, Cacioppo, & Rapson, 1994), the affect-as-information theory (Clore, Gasper, & Garvin, 2001) has gained increasing attention. Several studies have shown that the transmission of positive emotions from employees to customers leads to a better evaluation of the service experienced, as their current affective state serves as heuristic information (Pugh, 2001; Tsai & Huang, 2002). In this respect, however, the influence of emotional components in human-to-chatbot service interactions has not yet been studied. Thus,

while it is well documented from human-to-human research that customers evaluate service encounters based on more than functional aspects alone, this question is unanswered in the context of chatbot-handled encounters. This is problematic because it means that there is no comprehensive knowledge of the factors on which customer satisfaction in chatbot-handled service encounters is based.

Research gap 3: Can customers feel empathy toward a chatbot without any physical or graphical representation?

Third, the feeling of empathy has been the subject of various research efforts in the context of robots (e.g., Kwak, Kim, Kim, Shin, & Cho, 2013; Riek, Rabinowitch, Chakrabarti, & Robinson, 2009b; Rosenthal-von der Pütten, Krämer, Hoffmann, Sobieraj, & Eimler, 2013; Rosenthal-von der Pütten et al., 2014). Feeling empathy toward a social entity is based on anthropomorphism (Epley, Waytz, & Cacioppo, 2007), which, in contrast to the CASA paradigm, is characterized by a conscious attribution of human features. In the context of this thesis, results of extant research are subject to two major limitations: on the one hand, the feeling of empathy toward the robot has always been considered in the context of negative emotions. This means that the robot was either tortured (Rosenthal-von der Pütten et al., 2013; Rosenthal-von der Pütten et al., 2014) or bullied (Paiva, Dias, Sobral, Woods, & Hall, 2004) before assessing subjects' empathic reactions. On the other hand, compared to chatbots, robots possess a higher ability to trigger anthropomorphic reactions by humans due to their physical appearance (Duffy, 2003). The basic distinction between robots and chatbots (i.e., their physical appearance), therefore, raises the question of whether the physical appearance of an agent is a prerequisite for anthropomorphic reactions toward artificial entities. Thus, it is also necessary to investigate the feeling of empathy and anthropomorphic reactions toward chatbots in the context of positive emotions and without a physical appearance of the agent. The importance of addressing this research gap arises particularly from its interplay with the aforementioned CASA paradigm, as people at the conscious level normally negate the application of social behaviors toward artificial entities (Reeves & Nass, 1996). In this respect, the approach can create a deeper understanding of affective and social responses of humans toward chatbots, particularly in service encounters.

Research gap 4: Does a chatbot's graphical representation through an avatar affect customers' affective responses?

Fourth, looking at the implementation of chatbots in practice, it can be found that the representation through avatars, which are defined as "[…] computer-generated representations of the actors" (Bente, Rüggenberg, Krämer, & Eschenburg, 2008,

p. 292), is a very popular approach. In particular, the multitude of different implementation strategies, from corporate logos to cartoon-like human representations, shows that the mechanisms at this point are not yet fully understood. The extant research could show that the graphical representation of a conversational agent by a human-like avatar promotes anthropomorphic reactions (Lin, Doong, & Eisingerich, 2021) because the avatar serves as a human-like cue, which is used by interacting people to guide their own behavior (Blut, Wang, Wünderlich, & Brock, 2021). However, with regard to the use of avatars, the question remains unanswered if an avatar does influence customers' responses caused by a chatbot's human-like behavior through the expression of positive emotions (Felbo, Mislove, Søgaard, Rahwan, & Lehmann, 2017). An investigation of the research gap thus offers the potential to deepen the understanding of the effects of avatars in general, but especially regarding the interplay between the graphical representation of chatbots and their human-like behavior.

Research gap 5: Is a chatbot's affective delivery also beneficial in service recovery encounters?

Fifth, while chatbots are on the advance in the service sector, the extant research could repeatedly show that some customers exhibit reservations toward chatbots (e.g., Dietvorst, Simmons, & Massey, 2015; Mozafari, Weiger, & Hammerschmidt, 2020). The context of the service encounter seems to have a decisive influence on when these reservations are expressed. For example, Mozafari et al. (2020) reported a significant drop in customers' trust if revealed that they interacted with a chatbot when a critical service encounter was handled by the chatbot. Situations that are commonly considered as critical service encounters are service recoveries undertaken by organizations after the occurrence of a service failure (Johnston & Hewa, 1997). These service recoveries serve to rectify the inconvenience caused to the customer and are decisive for the continuation of the customer relationship (McCollough & Sundar, 1992). The expressed reservations raise the question of whether the use of chatbots, and thus also the possible favorable effects through their display of positive emotions, are limited only to routine activities. In addition, the use of emojis to substitute facially expressed emotions does not seem to be unproblematic in this context. For example, Thies, Menon, Magapu, Subramony, and O'Neill (2017) reported that the use of emojis or emoticons by customers was perceived as juvenile. In formal encounters aversions, in the sense of perceiving them as strange, could also be found (Duijst, 2017). The last research gap thus concerns the generalizability regarding emotional contagion triggered by a chatbot. Specifically, an answer is attempted to be found to the question of whether customers' affective responses to a chatbot's positive displayed emotion do occur also in service recoveries and if they are equally as beneficial as they are in routine encounters. An investigation of this research gap can provide important information on possible limitations regarding the use of affective chatbots, depending on the service context.

1.3 Objectives and Contributions

Objective 1: Emotional contagion triggered by chatbots

Six consecutive studies, which gradually strove to shed light on the outcomes of positive displayed emotions by chatbots, were conducted to close the outlined research gaps. Based on extant findings on the CASA paradigm (Reeves & Nass, 1996), *Study 1* strove to provide first ever empirical insights on how customers unconsciously react to a chatbot that displays positive emotions during the interaction and thus displays typical human behavior (Felbo et al., 2017). The expectation was that customers would unconsciously mimic the chatbot's displayed positive emotions as they would in human-to-human interactions. To test this assumption, a laboratory experiment was conducted, which relied on the Wizard of Oz method (i.e., a human operator secretly acts as chatbot). The unconscious changes in the customers' facial reactions were videotaped and then analyzed using automated facial coding software. The results of *Study 1* address the first research gap and partially close it by showing that the expression of positive emotions by a chatbot can trigger unconscious mimicry in customers. The results thus contribute to the literature on conversational agents and specifically on chatbots. They provide important evidence that the mechanisms (i.e., emotional contagion) known from human-to-human interactions can also be applied to human-to-chatbot interactions.

Objective 2: Relevance of emotional components for encounter satisfaction in chatbot-handled service encounters

The spread of chatbots in service is repeatedly motivated for by functional advantages such as permanent availability of the customer service (Schneider, 2017; Wünderlich & Paluch, 2017) or higher cost efficiency (Li et al., 2017). Because the extant research has shown that emotional components play an important role in human-to-human service encounters (e.g., Pugh, 2001), the second research gap outlined that in this respect, no knowledge exists regarding chatbot-handled service encounters. To address this research gap, *Study 2* and *Study 3* were conducted based on the affect-as-information theory (Clore et al., 2001). The studies pursued the objective to answer the question of whether a change in customers' affective state caused by the chatbot would lead customers to report higher satisfaction with the service experienced. By confirming this expectation, the results make a significant contribution to the service literature and show that chatbot-handled service encounters are by no means to be evaluated purely in functional terms. Instead, chatbot-handled encounters follow the mechanisms known from human-to-human research, where the service evaluation is influenced by emotional components. These results help in understanding more comprehensively the occurrence of satisfaction in chatbot-handled service encounters and are thus of importance for both scientists and practitioners. With this approach, the thesis follows a call for research approaches by Mustak, Salminen, Plé, and Wirtz (2021) to deepen the understanding of the mechanisms and modes of action of AI in service encounters.

Objective 3: Empathic reactions toward chatbots in service encounters
Based on theoretical insights from research on anthropomorphism, *Study 4* proposed empathy as a second mediation path that, in contrast to emotional contagion, is characterized by its more cognitive and conscious nature (Rogers, 1959). In human-to-human research, few research studies have examined empathy from the customer's perspective toward an employee (e.g., Wieseke, Geigenmüller, & Kraus, 2012). These found that empathy is an essential variable in the development and a determinant of the strength of the customer-employee relationship. In the context of conversational agents, the feeling of empathy toward an artificial agent implies anthropomorphism (Epley et al., 2007), the attribution of human features, such as an own affective state. Since existing research had so far investigated empathy toward artificial entities only in the context of negative emotions or using robots (e.g., Kwak et al., 2013; Riek et al., 2009b; Rosenthal-von der Pütten et al., 2013; Rosenthal-von der Pütten et al., 2014), there was a critical research gap in this regard. This was because it was not clear how the issue would behave in the context of positive emotions and applying a chatbot without any physical representation. It was therefore the third objective of the thesis to close this gap. The results show that displaying typical human behaviors in the form of positive emotions (Felbo et al., 2017) leads to triggering not only unconscious processes in customers (i.e., emotional contagion) but also conscious processes (i.e., customers more strongly anthropomorphize the chatbot and feel empathy). In the course of this extended investigation of the affective responses of customers to the positive emotions of a chatbot, the influences of different personality traits of customers were examined. Of particular interest were extraversion and openness to experience, acting as moderators. With the obtained results, the thesis contributes to both the service literature and the literature on conversational agents: first, the results show that the feeling of empathy toward chatbots also acts as a vital variable in chatbot-handled service encounters. Second, displaying typical human behavior bears the potential to cause anthropomorphic reactions

toward chatbots. Neither a physical nor a graphical representation is necessary for such reactions. Third, not all customers respond in the same way to the chatbot's affective delivery. It turns out that customers with a high level of extraversion appear more receptive to the chatbot's positive emotions. This shows that there should be a much stronger focus on individual traits of customers in research on chatbots, as this can significantly influence responses toward chatbots. The same applies to the use of chatbots in practice. While this can be beneficial in some customer segments, it may not have the same effect in others, and may even have negative consequences in some segments.

Objective 4: Effects of a chatbot's graphical representation on customers' affective responses

The extant research considers the presence of graphical human-like representations in the context of conversational agents, such as an avatar, as a vital factor concerning the attribution of human-like features (e.g., Corti & Gillespie, 2016; Lee, Kiesler, & Forlizzi, 2010; von der Pütten, Krämer, Gratch, & Kang, 2010). Previous research, however, has not been able to explain the influence of an avatar on affective customer responses to positive displayed emotions by a chatbot. Therefore, the fourth objective of this thesis was to investigate this moderating influence of an avatar. The results of *Study 5* show that simple graphical representations may affect the attribution of human-like features, but more so do human-like behavioral patterns such as displaying emotions. Thus, a vital contribution is made to the literature on conversational agents: that displaying more human-like cues seems to cause a decreasing marginal effect. In contrast, the effects caused by behavioral patterns may be inhibited by graphically designing a chatbot, so it contradicts the human-like behavior. On the one hand, this shows the importance of the chatbot's behavior during a service encounter. On the other hand, it shows that an unsuitably designed avatar can undermine the effects intended to be caused by the behavior. The results are important for literature on conversational agents. They show that a consideration of purely graphical implementations of chatbots falls short. Instead, behavioral features must be considered more intensively. In addition, the finding regarding the negative effects caused by a computer-like avatar is particularly relevant for practitioners, where the large number of different avatars gives the impression that there is no knowledge of the possible negative consequences.

Objective 5: Relevance of a chatbot's affective delivery in service recoveries

The reservations presented about chatbots in critical service encounters have strongly questioned whether a chatbot's affective delivery in the form of positive emotions can be beneficial in such encounters as well. It was the objective of *Study 6* to transfer the research context from a routine encounter to a more critical service recovery. With this, *Study 6* strove to extend the generalizability of the previous results. Furthermore, it should be explored if there are limitations regarding possible use cases for (affective) chatbots. By showing that emotional contagion between a chatbot and customers is not limited to routine encounters, the results contribute to the service literature and once more highlight the beneficial outcomes of enriching human-to-chatbot interactions by letting chatbots display positive emotions. This finding also has special relevance for practitioners. Since chatbots are not only resistant to catching the negative emotions of customers, but can also trigger a contagion of positive emotions, chatbots are ideally suited for use in service recoveries. Additionally, the results provide insights about how the successful handling of a recovery encounter affects the conscious process of empathy toward a chatbot.

1.4 Structure

The first chapter of the thesis, the "Introduction," began by highlighting the relevance of chatbots and their potential emotional performance in service encounters. Based on the progressive proliferation of chatbots in the service sector and the known relevance of emotions, the research gaps were outlined and the objectives highlighted. The central issue was that it has not yet been investigated how the display of positive emotions by a chatbot may affect chatbot-handled service encounters. An answer to this question is of paramount importance especially for the service literature because, in the context of emotional contagion, its influence on service-relevant outcomes such as encounter satisfaction is well documented. The "Introduction" concludes with an explanation of the structure of the present thesis.

For reasons of clarity, this thesis will organize the theoretical foundation into two main chapters that address the two central theoretical aspects. In these two chapters, first the foundations of AI and conversation agents are considered (Chapter 2) before addressing the investigated affective processes in the second half of the theoretical section. Since progress in AI technologies has been crucial for the progressive development of chatbots and for specific shifts regarding the research direction in recent years, the chapter begins with an introduction to AI (Section 2.1) before conversational agents are considered in detail (Section 2.2). There is a multitude of different approaches toward the topic of AI. To make it clear from which perspective this thesis approaches the topic, the different approaches are discussed before highlighting the relevant approach for the further course. Since research on chatbots, the text-based form of conversational agents, is also linked to text-based communication of emotions, this topic is addressed in the chapter as well. This chapter contains important information necessary to create a basic understanding, especially for the empirical part of the thesis.

Chapter 3 considers the second stream of research relevant for this thesis, emotions. The chapter begins with a consideration of the long-standing debate surrounding the topic of emotions and, more specifically, their definition (Section 3.1). The further course of the chapter then refers to the phenomenon of emotional contagion (Section 3.2). After the essential terms have been defined (Section 3.2.1), an examination of authenticity (Section 3.2.2) follows, which is derived and explained through the related topic of emotional labor. Authenticity was addressed via emotional labor because a suitable understanding of the concept of authenticity in the context of chatbots can be derived via the related constructs of deep and surface acting. The chapter concludes with a consideration of empathy (Section 3.2.3) in the further course understood as the cognitive counterpart to primitive emotional contagion, which will become part of the fourth study's research approach.

As explained above, the two preceding chapters refer to the theoretical foundations and are designed to clarify conceptual understandings. The present research approach combines various results from different streams of research. To give the reader a concise overview of the existing literature, which is central to the empirical part of the thesis, an overview of the current literature (Chapter 4) follows the theoretical foundations.

The empirical part of the thesis consists of six studies that build on each other. Since these studies have had different objectives and have been based on different methodological approaches, the beginning of the chapter provides a general overview of the studies (Section 5.1) to give the reader a compressed overview. Because the studies were based on different conceptual considerations and results, each subchapter of a study begins with a "Conceptual Development" where the hypotheses to be tested are developed. Since it is crucial for the further course of the discussion to know which objectives were associated with the respective study, a description of these is given as a second step in the chapter "Research Objectives and Setting." This is followed by a description of the stimuli used in the respective chapter "Experimental Design." Since pre-studies were conducted for some studies, this subchapter addresses them if applicable. The presentation of the studies' methodological aspects concludes with a description of the procedure ("Procedure") and the measurement instruments used ("Measures"). Before presenting the results, an outline of the participants is presented. The chapter "Results" is initially divided into evaluative test steps. Because common-method bias is a critical problem in research, several remedies were applied to address this issue. First, those remedies are discussed ("Common-method Bias"). As it is common procedure in experimental design, the results are in the second step checked concerning the experimental treatment ("Treatment Check") before the chapter "Path Model Estimation" explains the methodology used to calculate the proposed models and test the hypotheses. Before the hypotheses are tested ("Hypothesis Testing"), however, a detailed evaluation of the measurement model as well as a test of the structural model are performed. Since, as already explained, each of the studies carried out had its own objectives, the results obtained are discussed separately ("Discussion"). These respective chapters also serve as transitions between the studies that build upon each other. Some chapters can be omitted if, for example, a study did not include new measurement instruments compared to the previous one.

The "General Discussion" (Chapter 6) takes up the aspects already touched in the studies' discussion sections and reviews them in their entirety by bringing the results together. The thesis concludes with an outline of the implications and its limitations. This is divided into a discussion with a focus on scientific addressees within the chapter of the "Theoretical Implications" (Section 7.1) and a discussion designed to derive implications for practitioners (Section 7.2). The chapter proceeds with a discussion of the limitations and, based on this, an outline of future research approaches (Section 7.3) before the thesis concludes with a "Management Summary" (Section 7.4) to summarize key findings of this important topic for managers.

Foundations of Artificial Intelligence and Conversational Agents

2

2.1 Artificial Intelligence

2.1.1 Definition of Artificial Intelligence

Approaching the broad field of AI reveals a lot of different streams being interested in that topic, for example, computer sciences (e.g., Debauche et al., 2020; Ramos, Augusto, & Shapiro, 2008), health care (e.g., Esteva et al., 2017; Koch, 2018), economics (e.g., Heinen, Heuer, & Schautschick, 2017; Sion, 2018), and marketing sciences (e.g., Huang & Rust, 2018). All those disciplines have different parts of AI that are primarily important and being investigated. While, for example, marketing scientists mainly focus on use cases and increasingly on design principles, thus, mainly on the upsides of the use of AI (e.g., Lukanova & Ilieva, 2019), economists heavily focus on how the progress of AI will affect the labor market and possibly affects the loss of jobs (e.g., Davenport & Ronanki, 2018; Hamid, Smith, & Barzanji, 2017). Attention to AI, however, is not only given from a scientific perspective but also from a practical point of view (e.g., Kwidzinski, 2020; Sebes, 2018). It seems as if there is no business sector where AI is not one of the top priorities. As one would expect, practitioners do stress other aspects of AI, and it is often certain use cases that they discuss. This refers, for example, to voice-based agents being used in restaurants (Sebes, 2018). With those different angles looking at AI, many approaches in defining it have emerged over the years, and no universal definition has been yet agreed upon (Kaplan & Haenlein, 2019). In order to develop an understanding of why AI research is where it is today, it is crucial to illuminate the different approaches and to distinguish them from one another. This will also make it easier for the reader to understand the approach to AI taken by this thesis.

K. Prinz, *The Smiling Chatbot*,
https://doi.org/10.1007/978-3-658-40028-6_2

Russell and Norvig (2010) tried to structure different definitions and thus also different approaches to AI. They identified four different categories, from which two are based on rationality and two on human performance. Those two approaches to AI are distinct from each other in terms of their reference when assessing the performance of AI. While the "thinking rationally" as well as the "acting rationally" approach compare AI to rationality or what Russell and Norvig (2010) consider as the ideal performance, the approaches of "acting humanly" and "thinking humanly" compare AI's performance to human behavior or thinking. The latter approaches are in line with what is later introduced as the so-called Turing Test (Turing, 1950), a test to assess chatbot performance whether it is distinguishable from a human when communicating. Furthermore, the thinking humanly approach seeks to take the human brain and how it makes decisions as a basis. However, concerning the human approaches, Russell and Norvig (2010) point out that human behavior and thinking underlay irrational patterns, for example, influenced by emotion (De Martino, Kumaran, Seymour, & Dolan, 2006), as well as the fact that intelligence is not evenly distributed across all people. These facts may lead to a gap between the optimum and what is achieved when following the human-centered approach.

Based on the classification suggested by Russell and Norvig (2010), this thesis follows a human-centered approach. The reasons for this will become apparent in the further course of this paragraph. Following a human-centered approach does not mean that AI, as humans typically would, should make systematic errors, but that human behavior and human decision-making should be the basis for AI development. In line with this approach, for example, chatbots, but also other artificial entities, are increasingly seen as social actors when interacting with humans (Eyssel & Hegel, 2012; Lee, Peng, Jin, & Yan, 2006; Shum, He, & Li, 2018; Stock & Merkle, 2018; van Doorn et al., 2017; Wirtz et al., 2018). Therefore, it is essential to consider social aspects related to the interaction between AI and users in the development phase. A definition that supports that understanding is proposed by Bellman (1978) who claims, that AI is the ability of computers to perform"[…] activities that we associate with human thinking, activities such as decision making, problem solving, learning […]" (p. 3) by previously empirically observing and later transferring human thinking and behavioral patterns. This definition highlights what is later elaborated, for example, regarding the expression of emotions: in the development of AI applications, human behavior is increasingly the basis and starting point especially concerning aspects that determine those technologies' behavior. In addition, AI is more and more frequently not seen as a pure task performer anymore but as a social entity. This perspective regarding the understanding of AI is supported by several central pioneers for the knowledge about AI that exists today (e.g., Beal & Winston, 2009; McCarthy, 2007; Nilsson, 2005). This means, that they demand less focus on developing solutions for individual specific tasks but more focus on thinking, learning, and creating. They call their maxim human-level AI. The latest developments in the field of AI show that the approach based on human performance seems to be increasingly gaining acceptance on both the practical and the scientific side. This can be seen from the fact that more and more scholars today refer to the Turing Test (see Section 2.2.3) when approaching AI (e.g., Latah & Toker, 2018; Russell & Norvig, 2010). Furthermore, also from a practical perspective, the relevance of human characteristics increases. This can be concluded from an approach toward defining AI suggested by Amazon. They define AI as the attempt to replicate cognitive capacities, generally attributed to human intelligence, with tools from computer science (Amazon Web Services Inc.). This definition explicitly highlights human characteristics that are the basis for developing AI technologies.

2.1.2 Historical Development

What has already become clear is that research on AI is not a development happening in recent history. Related issues have been preoccupying scientists for decades. Under the prevailing understanding of AI today, the first work was published by McCulloch and Pitts (1943). In their work, they presented an artificial neural network for the first time, which was based on the so-called "all-or-none" law of nervous activity. One limitation of these early networks was they lacked the ability to learn (Schmidhuber, 2015). Formative for the use of the term AI, however, was John McCarthy, who in 1955 was in charge of organizing a workshop that took place at Dartmouth College (McCarthy, Minsky, Rochester, & Shannon, 2006). The efforts of the people involved during the summer of 1955 are today commonly considered as AI's date of birth (Latah & Toker, 2018; Russell & Norvig, 2010). Later, the first artificial networks were introduced that had the ability of either supervised (e.g., Rosenblatt, 1985) or unsupervised (e.g., Grossberg, 1969) learning. While supervised learning relies on input-output examples, these concrete examples are unnecessary for unsupervised learning (Russell & Norvig, 2010). The introduction of multilayer neural networks (e.g., Viglione, 1970) enabled utterly new possibilities, which, among other things, allowed the option of so-called deep learning, which is understood as a way of how AI is now able to learn autonomously (Genmod Project, 2016). Deep learning is algorithm-based and is used to, for example, recognize objects or understand language (Gentsch, 2018). The so-called layering implies neural networks being connected on several levels, which leads to greater efficiency (Alshamrani & Ma, 2019). LeCun, Bengio, and Hinton (2015, p. 436) describe deep learning as "[allowing] computational models that are composed of multiple processing layers to learn representations of data with multiple levels of abstraction." These named advances in deep learning have opened up new opportunities, especially in the area of conversational agents, which will be presented in detail later on (Shum et al., 2018). Especially concerning their communication skills, significant advances could be made due to deep learning. Literature discriminates between two possible ways to generate responses of conversational agents. They may be generated either retrieval-based (e.g., Yan, Song, & Wu, 2016) or generation-based (e.g., Serban et al., 2017). The first approach uses the wide variety of online available user inputs to generate an answer and thus uses extant human-to-human interactions (Higashinaka et al., 2016). Generation-based approaches are superior in that they can act detached from context constraints and enable more complexity regarding the conversation, as retrieval-based solutions do rest on predefined answers (Serban et al., 2017).

2.1.3 Levels of (Artificial) Intelligence

In addition to looking at the technical developments that enabled today's application options in the first place, especially for chatbots, it is also important to look at the intelligence aspect. This is important because it allows a better understanding of which associated stage of intelligence this thesis addresses and where the main challenges in the development of AI in the coming years are seen. Based on the development over time and concerning the resulting different levels of intelligence, Huang and Rust (2018) differentiate between four categories of AI. The basis is the definition of intelligence according to Gardner (1999, p. 33 f.) as "[…] biopsychological potential to process information that can be activated in a cultural setting to solve problems or create products that are of value in a culture." From this definition, the aspects of processing information and solving problems are of particular importance. If, as shown above, the human-centered approach (Russell & Norvig, 2010) is about mimicking human behavior, the de facto conclusion of the given definition is that the biopsychological process of intelligence is substituted by appropriate algorithms so that one can speak of intelligence in the mentioned sense. Based on the discussed definition, so-called mechanical intelligence can be found on the lowest intelligence level (Huang & Rust, 2018). For Huang and Rust (2018), this type of intelligence includes the ability to perform routine and repetitive tasks. One major discussion on AI that is heavily conducted, especially regarding lower-skilled jobs, concerns its impacts on jobs and those jobs that will potentially be replaced by AI (Huang & Rust, 2018). However, Huang, Rust, and Maksimovic (2019) assume it will be mostly on a task level where certainly replacements will happen and less regarding whole jobs. In this regard, the focus is particularly on those tasks that are characterized by regular and repeated exercise as well as having a low degree of complexity. The second stage is the so-called analytical intelligence (Huang & Rust, 2018). Sternberg (2005, p. 191) considers this type to be present when "[…] the information-processing components of intelligence are applied to analyze, evaluate, judge, or compare and contrast." However, he also points out that new problems must nevertheless have similarities to problems that are already known and were solved. The third level is the so-called intuitive intelligence (Huang & Rust, 2018). The central component of this form of intelligence, which comes close to human intelligence, is intuition. Intuition plays a decisive role in everyday life, although it is often associated with suboptimal and biased decisions (Lieberman, 2000). This is mainly because, in corresponding situations, not all available information may be taken into consideration and are outperformed by extant memory patterns (Broniarczyk & Alba, 1994). Intuition does also take a central position, especially during social interactions with other people, for example, with regard to displaying appropriate emotions (Barr & Kleck, 1995). For Sternberg (1999), intuitive intelligence refers also to dealing with new and previously unknown situations. Concerning AI, the transition from analytical to intuitive intelligence also means the transition from so-called weak AI to strong AI (Searle, 1980). Russell and Norvig (2010) suggest, as a distinction between weak and strong AI, that weak AI behaves as if it was intelligent, while strong AI is actually able to think and thus is intelligent. They emphasize that it is not about simulating thinking but about factual thinking. There have been different approaches to a definition of thinking (Arnold & Wade, 2015). In one approach suggested by Squires, Wade, Dominick, and Gelosh (2011), the possibility of weighing up and, to a certain extent, connected to that also the reduced limitation concerning specific contexts through the possibility of making connections, is stressed. For the handling of tasks, this transition furthermore implies that intuitive AI can learn from experience and, in particular, from mistakes (Huang & Rust, 2018). However, intuitive intelligence and AI that is capable of it still lacks one important ability. Human interaction is hardly conceivable without the influence of emotions (Ekman, 2000; Ekman, Freisen, & Ancoli, 1980; Myers, 1989). Emotional intelligence is widely considered as a crucial ability necessary for the interactions between humans (Schutte et al., 2001) and successful business environments (Fineman, 2004). This type of intelligence is the fourth and highest form of intelligence AI may reach. Salovey and Mayer (1990, p. 189) originally suggested the concept of emotional or empathetic intelligence. In their work, they define emotional intelligence as "[…] the ability to monitor one's own and others' feelings and emotions, to discriminate among them and to use this information to guide one's thinking and actions." For the present thesis, however, the definition mentioned above by Salovey and Mayer (1990) shows a significant weakness. It focuses too much on the perspective of the perception of emotions and thus, in the context of AI, takes too little into account the perspective of displaying emotions and possible outcomes of such behavior. In this respect, the definition by Mattingly and Kraiger (2019, p. 140) goes further as they state that emotional intelligence "[…] refers broadly to […] skills and/or abilities that enable awareness of the emotional states of oneself and others and the capacity to regulate or use emotions to positively affect role performance." Hatfield et al. (1994) argue in a similar direction about emotional intelligence. They emphasize the three characteristics of understanding the emotions of others, regulating one's own emotions, and using one's own emotions to trigger appropriate behavior, which, in combination, are understood as the essential components of emotional intelligence.

In the context of AI and emotional intelligence, however, there is an ongoing and pronounced debate in scientific research on whether AI can feel emotions (Huang & Rust, 2018). Participants of this debate split into two opposed camps: the representatives of one side come from the stream of philosophy and psychology. They support the point of view, that a physiological change is necessarily connected with feeling emotions (e.g., Lazarus, 1991). This point of view becomes clear by looking at a suggested definition of emotions by Keltner and Gross (1999, p. 468), who define emotions "[…] as episodic, relatively short-term, biologically based patterns of perception, experience, physiology, action, and communication that occur in response to specific physical and social challenges and opportunities." Suppose one chose to follow the definitions that emphasize that emotions are processes based on different neural systems (Ochsner & Gross, 2005), then the representatives of the attitude mentioned above must be agreed and AI would not feel anything like emotions. Arguing against this are representatives of the AI and computing perspective (e.g., Minsky, 2006; Picard, 1995). Supporters of this perspective argue that programming emotions is not different from programming cognition. If this point of view is further broken down, this means that emotions are primarily relevant in terms of their meaning and representation. Thus, it is less relevant if AI does have an emotional state, but it rather depends on their perception. Concerning emotional intelligence, Huang and Rust (2018, p. 159) propose a rather pragmatic approach by claiming that emotional intelligence would thus be existent if AI were able to at least behave as if it had an emotional state.

Comparing the literature on emotions in the context of AI, be it chatbots or robots, the computing perspective currently seems to be prevailing. This can be seen from the growing number of research projects that have dealt with AI technologies and humans interacting with them in the past few years. In these approaches, two questions are usually of particular interest: (1) emotional reactions of interacting people toward artificial entities (e.g., Kwak et al., 2013; Riek et al., 2009b; Rosenthal-von der Pütten et al., 2013; Rosenthal-von der Pütten et al., 2014) and (2) representation of emotions and other social components by artificially intelligent agents that are normally attributed to human-to human communication (e.g., Cassell & Thorisson, 1999; Klein, Moon, & Picard, 2002; Liao et al., 2018). The computing perspective (see above) finds support in the CASA paradigm (Reeves & Nass, 1996). In a series of experiments, Reeves and Nass (1996) found that when people interact with computers, they unconsciously exhibit social behavioral patterns. They concluded that people see computers as social interaction partners, which ultimately led to the paradigm's name. Since the introduction of this paradigm, it has formed the basis for many research projects in the areas of robotics (e.g., Kim, Park, & Sundar, 2013; Lee et al., 2006) or in a more specific context also conversational agents (e.g., Ho, Hancock, & Miner, 2018; Liao, Davis, Geyer, Muller, & Shami, 2016) and could be confirmed in these contexts (for a detailed treatment, please refer to Section 5.2.1). The so-called social response theory (Nass & Moon, 2000), which is associated with the CASA paradigm, explains why people unconsciously apply social rules when they interact with artificial entities. It states that even rudimentary signs of social behavior by, for example, computers can lead to people exhibiting social responses. Therefore, Nass and Moon (2000) suggest that it is the human-like appearance of artificial entities that is responsible for the social responses humans show when interacting with artificial entities. The shift toward a human-centered or a human-level AI in scientific research leads to an increasing shift toward considering the interaction between people and computers or representations of AI as social interaction. This follows the suggestion by Reeves and Nass (1996), and, in particular, shifts the focus away from specific technical questions to investigating how interactions can be enhanced (Biocca & Harms, 2002; van Doorn et al., 2017). The following chapter will take a closer look at the findings that have been made in this regard in recent years, with a particular focus on chatbots.

2.2 Conversational Agents

2.2.1 Definition of Conversational Agents

In most cases, people in their everyday life will experience AI in two forms: either by interacting with so-called service robots or by communicating with conversational agents. On the one hand, service robots typically perform physical tasks (Colby, Mithas, & Parasuraman, 2016). Extant literature delivers multiple approaches to define service robots. For example, Wirtz et al. (2018, p. 909) understand service robots as "[...] system-based autonomous and adaptable interfaces that interact, communicate and deliver service to an organization's customers." However, this definition lacks the physical aspect that is included in several studies. Furthermore, the term was commonly used when talking about service agents that performed a physical task and showed a physical representation (e.g., Giuliani et al., 2013; Han, Lin, & Song, 2012; Stock, 2018). This is in line with the definition provided by Colby and colleagues. They define service robots as "[…] technology that can perform physical tasks (e.g., driving, housework, serving in a restaurant), operate autonomously without needing instruction, and are directed by computers without help from people" (Colby et al., 2016, p. 5). Correctly, Jörling, Böhm, and Paluch (2019, p. 405) point out in their definition ("We define service robots as information technology in a physical embodiment, providing customized services by performing physical as well as nonphysical tasks with a high degree of autonomy.") that a physical task is not mandatory. An example would be a robot performing the check-in in a hotel. Currently, service robots are commonly attributed to what was described above as mechanical AI due to their limited abilities and use cases (Huang & Rust, 2018). An example of service robots being used in the frontline can be found from the German company Robotise that developed the robot Jeeves for the use in hotels (Kwidzinski, 2020). Jeeves replaces employees and can supply guests with beverages, food, and other items by bringing them to their rooms.

On the other hand, there is the group of so-called conversational agents, or sometimes also called virtual robots (Wirtz et al., 2018). The group of conversational agents includes, for example, the assistants known from smartphones and other smart devices such as Apple's Siri, Google Assistant, and Amazon's Alexa. For most people, conversational agents are the most apparent type of AI they get in contact with during everyday life. Compared to the mentioned service robots, conversational agents do not have any physical appearance, as they only exist digitally. The primary character is its participation in a dialogue with an interacting human (Jain et al., 2018). This makes the conversation the core of the delivered service. For this reason, Luger and Sellen (2016) use the term "dialogue system" when they refer to conversational agents. Over the last years, many synonyms have emerged to describe conversational agents (Astrid, Krämer, Gratch, & Kang, 2010). However, by taking a closer look at extant literature, it quickly becomes evident that many studies lack a clear distinction between what is considered as a service robot and what is referred to as a conversational agent. For example, Lee et al. (2010) use the term robot during their investigation to describe a situation at a reception. What they used for their study was an embodied conversational agent because a face represented it on a graphical interface (screen) while it interacted with customers through a voice-based interface. There was no physical task that was performed, as well as no physical representation of the agent. This gives evidence about the fluent transition between embodied conversational agents and service robots concerning used terms. One can conclude from this explanation that every service robot that at least occasionally communicates with customers can, at the same time, be considered as a conversational agent. Thus, service robots may be seen as a special type of conversational agents due to their additional physical representation and (sometimes) the performed task. In the further course, this thesis follows the definitions by Colby et al. (2016) and by Jörling et al. (2019) suggested above that outlined the core aspects, especially the physical representation, that defines service robots and thus distinguishes them from conversational agents in the narrow sense.

2.2.2 Communication Channels

Conversational agents can interact with users via different communication channels. These different channels offer the possibility to further separate conversational agents into different categories. Basically, there are two possible ways for conversational agents to interact with people: through voice-based or text-based communication. Sometimes, conversational agents can additionally be separated, depending on the existence of a graphical representation. Wirtz et al. (2018) refer to this category as video-based agents. Other authors (e.g., Jain et al., 2018) follow the categorization of conversational agents based on their communication channel. However, they refer to them, if they are graphically represented, as multimodal embodied conversational agents.

Voice-based conversational agents

An increasing number of people use voice-based, also referred to as speech-based (e.g., Jain et al., 2018), conversational agents like Apple's Siri or Amazon's Alexa

(Wirtz et al., 2018). This type of conversational agent interacts with users using spoken language input and output. Those mentioned examples of voice-based conversational agents are often referred to as (intelligent) personal assistants (e.g., Sarikaya, 2017). Voice-based conversational agents are commonly considered offering a greater extent of human-like behavior as they can, for example, modulate their voice pitch or tone (Luo, Tong, Fang, & Qu, 2019). However, the extant research shows that conversational agents still appear to have limited capacities from the perspective of conversational agent users, especially concerning complex tasks, and thus are primarily used to fulfill simpler tasks (Luger & Sellen, 2016). In addition, anyone who has used Apple's Siri or Amazon's Alexa knows that their capabilities currently are still far from matching the mentioned benefits around voice pitch and tone.

Text-based conversational agents

Concerning text-based conversational agents, the discrimination of different intelligence levels needs to be considered (see Section 2.1.3). This is because conversational agents and especially text-based conversational agents do often include self-service technologies (SST) that attracted a lot of attention of scholars (e.g., Meuter & Bitner, 1998) and practitioners (e.g., Karmarkar, 2004) in the past years. Based on the presented levels of intelligence, SST typically have to be placed at the bottom of the scale, performing tasks that can be attributed to the level of mechanical intelligence. This is because they belong to the category of so-called task-completion systems developed to fulfill specific and clearly limited tasks (Shum et al., 2018). One of the main challenges in text-based conversational agents, especially when belonging to the group of SST, is their limited potential to participate in a social interaction as they are mostly developed to fulfill simple goal-related tasks (van Doorn et al., 2017). This distinction shows that text-based conversational agents cover a wide range of application areas. Depending on their intelligence and the associated capabilities, they range from SSTs on the one hand to chatbots on the other, as will become clear in the further course of this thesis. While performing functional tasks is in the foreground for the former, the question of how these tasks are performed in the sense of a human-like interaction is increasingly coming to the fore for the latter (Jain et al., 2018). In short, chatbots offer a more natural conversation while the handling of tasks is in the foreground with SST.

Multimodal (embodied) conversational agents

When following Jain et al. (2018), multimodal embodied conversational agents must be considered a third subgroup. As an addition to text-based and speech-based conversational agents, they consider agents as belonging to that group if conversational agents do have a graphical representation that can display human-like behavior such as gestures (Bickmore & Cassell, 2005; Kopp, Gesellensetter, Krämer, & Wachsmuth, 2005). This approach indicates that it is primarily (human-like) graphical representations that are being used in designing conversational agents to enhance their appearance and trigger social behavior of interacting humans (Corti & Gillespie, 2016; Lee et al., 2010; von der Pütten et al., 2010). In line with these findings, scholars started shifting their attention to what is called social presence (e.g., Verhagen, Van Nes, Feldberg, & Van Dolen, 2014). In the context of social presence and artificial agents, some scholars also speak of so-called automated social presence (e.g., van Doorn et al., 2017). The question of social presence or automated social presence was raised because of the desire to enhance interactions between humans and conversational agents through their representation. Social presence in the context of conversational agents was defined by (Lee, 2004, p. 45) as "[…] a psychological state in which virtual (para-authentic or artificial) social actors are experienced as actual social actors in either sensory or nonsensory ways." Research found that representing a conversational agent with an avatar (i.e., "[…] computer-generated representations of the actors" (Bente et al., 2008, p. 292)) enhances the feeling of being with another social actor (Angga, Fachri, Elevanita, & Agushinta, 2015). This research stream emphasizes that research around conversational agents increasingly focuses on making the interaction between conversational agents and humans more human-like and consider conversational agents as social actors.

2.2.3 Chatbots as Text-based Conversational Agents

Definition of chatbots

Chatbots belong to the most used types of text-based conversational agents and AI in general (Oracle, 2016). Their relevance becomes clear by looking at the number of around 300,000 chatbots used in Facebook Messenger in 2018 (Boiteux, 2019). In the past years, the development of chatbots and the underlying functions attracted a lot of attention from both scientists, as will become apparent in the further course, and practitioners (Steger, 2017). As explained in the discussion on conversational agents, several names were used to describe chatbots, such as multimodal agents and conversational interfaces (Chaves & Gerosa, 2021). In the literature, chatbots are also referred to as text-based conversational agents (e.g., Gnewuch, Adam, Morana, & Maedche, 2018), chatter bots (e.g., Chakrabarti & Luger, 2015; Chaves & Gerosa, 2021), or social bots (e.g., Murgia, Janssens, Demeyer, & Vasilescu, 2016). Liao et al. (2018) highlight their text-based communication and their disembodied modalities as central characteristics. This approach supports what was already stated earlier: conversational agents, including chatbots, have to be distinguished from service robots based on their missing physical appearance. Because of that, some authors do also refer to chatbots as disembodied agents (e.g., Araujo, 2018). However, what is missing in the suggestion by Liao et al. (2018) shown above, is the reference to human-to-human communication. Thus, Crutzen et al. (2011, p. 514) present a more useful definition for the further course of this thesis by defining a chatbot as "[…] an artificially intelligent chat agent that simulates human-like conversation, for example, by allowing users to type questions (i.e., queries) and, in return, generating meaningful answers to those questions." Furthermore, chatbots are characterized by their turn-based messaging (Jain et al., 2018), meaning that their design is oriented toward the nowadays commonly used type of texting via messenger applications such as WhatsApp using relatively short messages and many turns compared to rather long messages and fewer turns as is typically known from e-mails. McTear et al. (2016) argue in the same direction and outline that it is a central characteristic of chatbots that people interact with them as if they were interacting with humans. This aspect is a decisive reason why companies replace employees with chatbots, which in the further course leads to chatbots more and more forming the interface between a company and its customers (Chaves & Gerosa, 2021).

Historical developments

Developments in the field of AI have also led to more attention being paid to chatbots in recent years (Riikkinen et al., 2018). However, research on chatbots goes back to the 1950s. Alan Turing (1950) was a pioneer in this field. He developed the so-called Turing Test to examine whether people can distinguish a chatbot from a human interaction partner. The key message was that AI (i.e., an artificially intelligent chatbot) would have to be present if a human would no longer be able to distinguish a chatbot's answer from that of a human (Dahm & Dregger, 2019). The first developments around the topic of chatbots were primarily about passing the mentioned Turing Test (Shum et al., 2018). For this reason, these developments were structured so that they acted in a limited field of activity. In 1966, the Massachusetts Institute of Technology developed the first chatbot, called ELIZA (Weizenbaum, 1966). However, the first chatbot to pass the Turing Test was the Parry chatbot developed by Colby (1975). In particular, due to fundamental developments in the field of AI, the capabilities of chatbots have been greatly expanded in the period since ELIZA and Parry (Shum et al., 2018). One of the most notable examples of recent developments is the XiaoIce chatbot, developed by Microsoft. This chatbot and its capabilities are a noteworthy example of the progress AI has made and which possibilities those advances offer for related applications. XiaoIce furthermore is designed to not being limited to the performance as a chatbot but also being able to, for example, draw pictures (Spencer, 2018).

Current research focus

For many years, research around chatbots has primarily focused on the technical aspects (Jain et al., 2018). With the increasing use of chatbots due to the new technological possibilities, the question regarding their design aspects increasingly received attention from a scientific perspective. This means that increasingly issues around their graphical appearance or their behaviors during communication are observed. Thus, in the past years, scholars conducted many studies on design principles, such as, for example, their behavior (e.g., Jain et al., 2018; Liao et al., 2018). One question was and is being investigated in particular: should, and if so, in what way, chatbots disclose their nature during a customer interaction. Concerning disclosure, research shows ambivalent results. On the one hand, Jain et al. (2018) showed that especially first-time users appreciate when the chatbot reveals its nature and at the same time provides information about what its capabilities are. On the other hand, research showed that disclosing the chatbot's nature may also negatively affect perceived service performance since customers exhibit reservations toward chatbots (e.g., Luo et al., 2019; Mozafari et al., 2020). In this context, research has shown negative effects of disclosure on general acceptance (Murgia et al., 2016) as well as on persuasion (Shi et al., 2020). Luo et al. (2019) found that customers exhibited rougher behavior while interacting and made fewer purchases in a marketing context. These results came about even though the pure performance of a chatbot was comparable to that of an experienced employee and exceeded that of an inexperienced employee by four times. For their study, they used a voice-based chatbot but pointed out that those results will likely be applicable also for text-based chatbots. Mozafari et al. (2020) have recently shown that trust plays a central role in the context of chatbot disclosure. In general, the conclusion must be drawn that, although negative effects are to be expected from a disclosure, an explicit and direct disclosure of a chatbot's nature in customer interactions should be made (Schanke et al., 2021).

Another field of research concerns the answers of chatbots. In this regard, research moves in a tension between overly short answers delivering an insufficient quantity of information (Jenkins, Churchill, Cox, & Smith, 2007) and possible too detailed responses to simple inquiries (Gnewuch, Morana, & Maedche, 2017). From a scientific perspective, investigations on designing chatbot answers no longer focus solely on purely technical aspects (i.e., how the answers are generated). Instead, there is broad consensus that chatbots, in addition to meeting the basic requirements associated with the respective service encounter, now have to be developed, so they conform to social requirements that are usually present during human-to-human interactions (Jain et al., 2018; Jenkins et al., 2007; Thies et al., 2017). The underlying assumption is that the acceptance and thus the effectiveness of chatbots and other conversational agents can be enhanced by making them fit more naturally into interactions with humans by focusing on human needs (Dautenhahn, 1999). Wallis and Norling (2005, p. 35) subsume this: "[…] to be an acceptable interface, a conversational agent must be accepted as a social actor, must display social intelligence and must participate in the social hierarchy." Wirtz et al. (2018, p. 915), in a robot context, argue in the same direction by stating: "In sum, it seems reasonable to assume that consumer acceptance of service robots depends on how well robots can deliver on the functional needs (i.e., related to dominance) and the social-emotional and relational needs (i.e., related to warmth) to achieve role congruency." Thus, it is crucial to consider that by replacing more and more frontline employees, chatbots are increasingly assigned the role of a representative of the respective company (Larivière et al., 2017; Marinova et al., 2017; Mozafari et al., 2020; Verhagen et al., 2014). For this reason, it is crucial that, as one would train employees regarding company requirements (e.g., regarding requested behavior), those requirements are also taken into consideration when chatbots are designed to be used in this context. In this regard, Morrissey and Kirakowski (2013, p. 95) state that chatbots "[…] should use appropriate grammar and spelling consistently, and consistently adopt an appropriate linguistic register with the user." In terms of customer interactions and, in a broader context, service encounters, scholars could also show that customers associate gender stereotypes with specific services and thus even with chatbots if they perform those services. For example, Forlizzi, Zimmerman, Mancuso, and Kwak (2007) showed that female representations of a conversational agent were preferred for potential tasks traditionally carried out by women. These included, for example, librarians and matchmakers. In contrast, male representations were preferred for tasks such as that of an athletic trainer. Since participants in both subsequent studies consistently stated that they preferred human-like representations, the authors concluded that consideration of gender must play a critical role depending on the intended area of application. Aspects such as the behavior in interactions, gender, and personality belong to the chatbot's identity, whose consistent presentation is considered an essential characteristic (Bartneck, 2002; Shum et al., 2018). Regarding behavioral patterns, scholars found that people tend to like artificial agents more if their personality is similar to their own (Jung, Lim, Kwak, & Biocca, 2012). All aspects of a chatbot, including its personality, are things that a chatbot does not define itself, but where it depends on the work of the programmers (Chaves & Gerosa, 2021). In this respect, it is essential that findings from scientific research are transferred to the practical development and application of chatbots.

2.2.4 Emotions in Text-based Communication

For humans, emotions are an essential part of life and vital in social interactions (Ekman, 2000; Ekman et al., 1980; Myers, 1989). From this aspect, some scientists conclude that it is a central requirement for chatbots to perceive and express emotions to pass the Turing Test (Picard, 1995). However, there is the widespread opinion that text-based communication bears the special challenge of communicating social information, including emotions (Byron, 2008; Huang, Yen, & Zhang, 2008; Kiesler, Siegel, & McGuire, 1984; Walther & D'Addario, 2001). Although this question would have been relevant in the context of chatbot research, it has long been neglected due to a strong technical focus for many years. With the advent of e-mail communication in the 1990s, this focus changed and research firstly analyzed the use of emoticons in human-to-human communication (e.g., Rezabek & Cochenour, 1998; Walther & D'Addario, 2001). According to Rezabek and Cochenour (1998, p. 201), emoticons can be understood as "[…] visual cues formed from ordinary typographical symbols that when read sideways represent feelings or emotions". While research in this regard has been and continues to be focused on human-to-human contexts, it is important to note that it is also an important foundation for designing interactions with conversational agents. In this context, it must be pointed out that the use of emoticons (and later also emojis) mostly is a substitute for facially displayed emotions. However, while facially expressed emotions are considered as nonverbal behavior, emoticons and emojis are not (Walther & D'Addario, 2001). Besides the expression of emotional states, emoticons can express also voice inflections and even information on body language in general (Mackiewicz, 2003). In a laboratory study, Churches, Nicholls, Thiessen, Kohler, and Keage (2014) investigated in which areas of the brain emoticons perceived by a recipient are processed compared to perceived facial expressions. Remarkably, they found indications that in both cases the processing (i.e., the understanding) takes place primarily in occipitotemporal sites. In contrast, however, it seemed that the perception of emoticons takes place in different areas compared to perceiving facially expressed emotions because perceiving emoticons did not activate the laterally placed facial feature detection systems. In accordance with Jeffreys (1993), who investigated human faces, the authors concluded that emoticons' perception and processing are learned patterns (Churches et al., 2014).

However, emoticons are only one possible way of expressing emotions in text-based communication (Felbo et al., 2017). In recent years, in particular, emojis, colored pictograms that are not rotated but upright (Ganster et al., 2012), have established themselves as an additional, if not the primary, way of expressing emotions. The term emojis comes from Japanese (Danesi, 2016). The "e" stands for "picture" and the "moji" for "letter." Thus, Danesi (2016, p. 2) concludes that the meaning of the word is "picture-word." The Oxford Dictionary suggests defining an emoji as "a small digital image used to express an idea or emotion in e-mails, on the internet, on social media, etc." (Oxford Dictionary) Sometimes emojis are also referred to as smileys (e.g., Lohmann, Pyka, & Zanger, 2017). Besides the advancing use of e-mails, two further developments give emojis as a substitute for emotions and the associated research an additional boost: on the one hand, communication via so-called messengers (Yoon, Lee, Lee, & Lee, 2014) is establishing itself today in addition to classic communication via e-mails. The increasing relevance of this form of communication is shown, among other things, by the fact that customers seem to report a higher willingness to write a short message compared to calling for a service request (Fowler, 2018). On the other hand, today's devices offer a wide range of options via emojis to enrich the own statements concerning their emotional content. These two mentioned developments around the widespread use of smartphones and the associated possibilities have led to emojis becoming established in today's communication (Felbo et al., 2017). How well established these expressions are, is shown by the fact that in 2015 the emoji "face with tears of joy" (😂) (Emojipedia, n.d.-a) was chosen as word of the year by the Oxford Dictionary. The devices used for communication today offer a barely tangible number of emojis. For example, only the update to iOS 14.2 for Apple devices brought over 100 new emojis to iPhones (support.apple.com/de-de/HT211808). One would expect that a high number of redundancies can be found in this large number. However, the meaning is often slightly different (Felbo et al., 2017), which makes the use a problematic matter in many cases. Official lists and tables can be used as a basis for the interpretation and the differentiation of the various emojis (Eisner, Rocktäschel, Augenstein, Bošnjak, & Riedel, 2016). In general, when it comes to communicating emotions, it can be stated that emojis (68.1%) were predominantly used in the available studies and emoticons played a relatively minor role (30.9%) (Vidal, Ares, & Jaeger, 2016). In this regard, Vidal et al. (2016) also found that emojis are more likely to be used to express positive emotions (66.7%).

However, there appears to be a main restriction regarding the use of emojis and emoticons in service. In contrast to the display of emotions in regular face-to-face communication (Bartneck, 2002), the use of emoticons or emojis in text-based communication is sometimes described as informal (Lebovits, 2015), juvenile (Thies et al., 2017), or, during situations with a high degree of formality and urgency (e.g., blocking a stolen credit card), viewed as strange (Duijst, 2017). Other studies conclude that the use of emotional substitutes harms perceived competence (Li, Chan, & Kim, 2019). However, Li et al. (2019) furthermore point out that there are significant differences between more exchange-oriented customers and customers with a communal relationship orientation. In sum, the following must be considered. Despite the widespread attention research pays to emojis, there are still blank spots and not all questions have finally been clarified. For this reason, it is not possible to say with certainty, regarding the previously mentioned research results, to what extent the negative reactions are due to the use of emojis or rather the general expression of emotions by the employees.

From the previous explanations, it becomes clear how important emojis are in today's communication. They are probably the most apparent method of transmitting emotions in computer-mediated communication. However, it must be taken into consideration that the communicated content of the emojis is only one side of the coin regarding the emotional content of the overall message. Apart from the research on either emoticons or emojis, researchers are also looking into the question of how emotions are expressed purely text-based (i.e., without any additional graphical representation) (Gill, Gergle, French, & Oberlander, 2008). While one generally speaks of explicit communication when using emoticons or emojis, the language-based presentation refers to implicit communication (Derks, Fischer, & Bos, 2008). In the context of computer-mediated communication, Siegel, Dubrovsky, Kiesler, and McGuire (1986) determined that this form of communication promotes the occurrence of so-called flaming, typically characterized by offensive behavior (Moor, Heuvelman, & Verleur, 2010). In their experiment, Siegel et al. (1986) showed that test subjects who exchanged information through computer-mediated communication were more likely to use strong language, expressing negative emotions. Although research approaches have tried to shed light on flaming in computer-mediated communication, the question remains as to what the drivers are for causing flaming to occur with a higher frequency in computer-mediated communication than in face-to-face communication (Derks et al., 2008). The same possibility of directly communicating emotional states exists also in a positive direction (Byron, 2008). In addition to expressing emotions via content, there is also the possibility of computer-mediatedly conveying emotional components via tone of the voice. In a laboratory experiment, Sasaki and Ohbuchi (1999) let people interact with a confederate either via computer-mediated communication or via a voice channel. They found evidence for people being able to communicate and interpret emotions through computer-mediated communication based on the tonality of the text. The results were supported by Hancock, Landrigan, and Silver (2007). During an experiment, they found that increased agreement, less use of negatively valenced terms, the use of exclamation marks, and longer phrases are indicators for the expression of positive emotions in text-based communication.

Emotions and Their Relevance for Service Research

3

3.1 Emotions

The importance of emotions for people's social interaction is undisputed (Ekman, 2000; Ekman et al., 1980; Myers, 1989). Also, their relevance in marketing contexts, for example, brand management (e.g., Kenning et al., 2005; McClure et al., 2004) or service management (e.g., Grandey et al., 2005; Pugh, 2001) was proven by several studies. Research paid additional attention to emotions due to new technologies like functional magnetic resonance imaging (fMRI) (e.g., Stoll, Baecke, & Kenning, 2008) or AI-enabled solutions to track facially displayed emotions (e.g., Smith & Rose, 2020). While there is consent on their relevance, there is still a debate on what emotions are and how they can be defined (Hatfield, Cacioppo, & Rapson, 1993). Ochsner and Gross (2005, p. 242) suggest a possible approach that defines emotions as "[…] valenced responses to external stimuli and/or internal mental representations […]". In their definitional approach, they stress that these responses are based on different response systems that may be activated due to emotional experiences. For Ochsner and Gross (2005) emotions are further characterized by their dependency upon multiple neural systems involved in their processing. These systems, responsible for the detection and classification of a stimulus as emotional and the resulting generation of an affective state, are the ventral system and the dorsal system (Phillips, Drevets, Rauch, & Lane, 2003). Furthermore, Ochsner and Gross (2005) draw a clear line between emotions and moods. They consider emotions as both unlearned and learned responses, for example, due to conditioning. The essential aspects of Ochsner's and Gross's (2005) approach, which are central for the further course of this thesis, will be examined in more detail below.

K. Prinz, *The Smiling Chatbot*,
https://doi.org/10.1007/978-3-658-40028-6_3

First, emotions result from external impact factors (Ochsner & Gross, 2005). This means, feeling a specific emotion is always preceded by a related event that causes this emotion. This is an essential aspect, as scientific literature does not always distinguish between emotions and so-called mood. While, as explained above, emotions are short-termed intense feelings, moods are considered to be less strong but therefore longer-lasting (Tellegen, 1985). Furthermore, the distinction between moods and emotions contains also what causes them. While it is a specific stimulus that triggers an emotion (Ochsner & Gross, 2005), the origin of a mood is generally not clearly identifiable (Fisher, 1997; Ochsner & Gross, 2005). What is important to mention, especially regarding the later introduced concept of emotional contagion, is that emotions contain an awareness component and include conscious notice of the person experiencing them (Hatfield et al., 1993; Hatfield, Carpenter, & Rapson, 2014). This means, if asked, a person would, for example, be able to explain the current affective state and probably be able to mention the reason for it. Concerning different terms that are being used, it must be made clear that research often does not clearly discriminate between emotion and affect. This becomes obvious, especially in English literature (e.g., Barsade, 2002; Hatfield et al., 1994). For example, Barsade (2002, p. 646) states: "I use the term emotion in this paper as a broad label, similar to that of 'affect,' both of which interchangeably encompass the general phenomenon of subjective feelings." For the further course, this thesis will refer to the affective state as a synonym to emotions, considering it the emotional response to the interaction with the chatbot and the affective delivery (i.e., the display of positive emotions) of that chatbot during the empirical investigations.

The second important aspect the definition presented above emphasizes is the valenced response to the mentioned stimuli (Ochsner & Gross, 2005). Valence means that emotions are typically categorized as being either positive (i.e., pleasant), negative (i.e., unpleasant), or neutral (Byron, 2008). In this context, neutral refers to a situation where neither positive nor negative emotions are present. Besides their valence, also their intensity, which is regularly referred to as arousal, determines the experienced emotion. Emotional arousal refers to the level of activation associated with a specific present emotion (Russell, 1980). The concept around arousal becomes clearer by looking at an example provided by Byron (2008). She (p. 310) describes emotions' arousal as follows: "[…] rage and elation are higher in intensity than frustration and contentment […]." This means, while, for example, elation and contentment are both positively valenced emotions, the arousal is higher for elation, making it a stronger felt positive emotion than contentment.

Another central topic in emotion research was the distinction between basic and non-basic emotions (Fisher, 1997). Basic emotions are considered to be emotions that can be found universally and are understood across cultural boundaries (Izard, 1977). Over the years, various theories emerged with the number of identified basic emotions ranging from two (Mowrer, 1960) to 17 (Frijda, 1986). For some scholars, the existence of basic emotions implies that all other possible emotional states are combinations of the primary emotions (Plutchik, 1962). Ekman, Levenson, and Friesen (1983) suggested one of the most commonly known approaches. They were able to show that different reactions in the autonomic nervous system were triggered when certain emotions are induced by facially displaying related emotions. In total, they tested six different emotions (surprise, disgust, sadness, anger, fear, and happiness) that also led to feeling the emotions when participants were told to display them facially. Another approach forwarded by Ekman was named the so-called facial action coding system (FACS), as they connected facial reactions with specific emotional states (Ekman et al., 1980). They based their theory on the assumption that specific muscle activity patterns in the face are related to feeling certain emotions (Hatfield et al., 1992). Based on the FACS, Ekman et al. (1980) concluded, for example, that smiling must not necessarily be a display of positive emotions but does so if the zygomatic major muscle is involved. As will become evident during the empirical investigation (*Study 1*), the concept of basic emotions and the FACS is commonly used among practitioners, for example, to reduce the otherwise high complexity in the measurement of emotions to a more understandable measure. The resulting measurement tools are becoming increasingly popular among scientists (e.g., Smith & Rose, 2020).

3.2 Emotional Contagion

3.2.1 Definition of Emotional Contagion

Everyone has experienced at least once the phenomenon of interacting with another person and somehow feeling cheered up afterwards. What has happened is the transfer of emotions from one person to another, called emotional contagion (Hatfield et al., 1992). The way displayed emotions affect people's emotions around has caught researchers' interest since decades. The term was first introduced by Scheler (1948, p. 11), who talked about it as *Gefühlsansteckung*. Early researches assumed that emotional contagion is a process people have to learn (e.g., Aronfreed, 1970). Other scholars thought of emotional contagion as a complex cognitive process that leads to people feeling the same as those around them (Hatfield et al., 1992).

Nowadays, however, there is a clear distinction between catching another person's emotions based on a sophisticated cognitive process or somewhat unintentionally and unconsciously (Hatfield et al., 2014). There is a consensus about the cognitive process that discriminates conscious empathy from primitive emotional contagion. Primitive emotional contagion is seen as "[...] the tendency to automatically mimic and synchronize movements, expressions, postures, and vocalizations with those of another person and, consequently, to converge emotionally" (Hatfield et al., 1992, p. 153 f.). Emotional contagion is not an end in itself: the contagion serves to process another person's affective state by practically feeling the state oneself (Prochazkova & Kret, 2017). In contrast to empathy (see Section 3.2.3), primitive emotional contagion happens primarily unconsciously and unintentionally (Hatfield et al., 1992) and during an interaction in a relatively short time (Du, Fan, & Feng, 2011). For reasons of simplicity, the term emotional contagion is used in the further course of this thesis when referring to primitive emotional contagion and thus talking about the unconscious processes. If the conscious process, empathy, is addressed, it will be explicitly mentioned and coined as such.

When interacting with other social beings, everyone is continually exposed to surrounding people's emotions expressed primarily by nonverbal signals such as facial expressions, postures, voices, and movements (Hatfield et al., 1992). Being exposed to these emotional expressions, the actual emotional contagion is triggered by the automatic and unconscious mimicry of those expressions (Hatfield et al., 1992; Hatfield et al., 1993; Hatfield et al., 1994). The process of primitive emotional contagion happens in three steps. In the first step, there is mimicry of the interacting counterpart's emotions by the subject (Barsade, 2002). As mentioned above, this mimicry happens unintentionally and unconsciously (Hatfield et al., 1993). Subsequently, in a second step, a feedback process occurs, as mimicking the counterpart affects the subject's affective state, which ultimately leads to a consciously reportable affective state, the actual contagion (Barsade, 2002). While steps one and two happen on an unconscious level, step three is the one that, for example, if asked for the current affective state, reaches a conscious level again (Barsade, 2002; Hatfield et al., 1993). The various communication channels through which both mimicry and feedback can occur are addressed below.

Vocal mimicry and feedback

One important channel for the contagion process of emotions in personal communication is the voice. Evidence for the existence of mimicry of speech was already found in the early 1980s by Condon (1982). In this time, investigations revealed people's ability to mimic and synchronize speech with their counterparts within one-twentieth of a second (i.e., unconsciously). Furthermore, Ekman, Friesen, and Scherer (1976) found a difference in pitch as they elicited certain positive or negative emotions in subjects and presented audio recordings to other subjects who were supposed to rate the emotional state. In this context, Hall (2005) supported the vocal feedback hypothesis with her findings. She could show that participating subjects' emotional state was significantly impacted by how they had to reproduce specific sound patterns (e.g., joyous or sad). Subjects who had to reproduce a joyous speech pattern reported significantly higher joy scores than subjects from other conditions (e.g., fear). It remained unclear whether it was hearing the sound patterns or reproducing them themselves instead that mainly caused the change in the emotional state. One of the first to find an answer to this question were Hietanen, Surakka, and Linnankoski (1998). They could link a facial reaction to a vocal expression with either a positive or negative valence and, with that, provided evidence for the existence of emotional contagion based on vocal expressions. Their findings were later supported by Neumann and Strack (2000), who exposed subjects to speech patterns with either a positive, negative, or neutral valence. After the exposure, they measured the subjects' affective state and found a significant impact from the speech pattern on the affective state, thus supporting the vocal mimicry and feedback theory.

Postural mimicry and feedback

Besides speech patterns, posture, as well as body language, may play an essential role in the contagion process of emotions. Sometimes, postural mimicry is also referred to as motor mimicry (e.g., Chartrand & Bargh, 1999; Prochazkova & Kret, 2017). That there is some kind of postural mimicry was first stated by Scheflen (1964). He observed several social encounters and claimed that there is a tendency for congruency to appear. He specified this to happen when the postures are synchronized. Later, Bavelas, Black, Lemery, and Mullett (1986) investigated people mimicking expressed emotions such as smiling or disgust. Later studies by Bernieri (1988) as well as Chartrand and Bargh (1999) could confirm those results by either investigating it in a student-teacher relationship (Bernieri, 1988) or by letting two participants interact while being given the task to describe photographs (Chartrand & Bargh, 1999). While both studies confirmed the occurrence of mimicry, Bernieri (1988) found a positive impact on the rapport between the two interactants. Chartrand and

Bargh (1999) found participants to report higher liking toward their interaction partner and a smoother run of the interaction. Support for the feedback hypothesis stems from Duclos et al. (1989). During their experiment, they told participants to adopt sitting positions and thus postures associated with fear, sadness, or anger. After resting in this position for at least 15 seconds, participants were asked for their affective state. They reported significantly more of that kind of affect intended to be triggered by the corresponding posture. With this, they reported evidence for the feedback process through postural movements.

Facial mimicry and feedback

Possibly the most important and prominent channel for emotional contagion is facial expressions of emotions. This is due to the high number of muscles in the human face, enabling a high degree of expressiveness (Prochazkova & Kret, 2017). Several studies have investigated facial mimicry as well as the feedback caused by facially displayed emotions. Early support for facial mimicry taking place was provided by McHugo, Lanzetta, Sullivan, Masters, and Englis (1985). They exposed participants to images showing either smiling or frowning faces. They reported strong evidence for participants mimicking those corresponding faces, which affected the affective state based on self-report scales. Later, Dimberg, Thunberg, and Grunedal (2002) conducted a laboratory study that included one configuration where participants were requested to consciously avoid reacting to shown happy faces. During the experiment, they measured the activity of the zygomatic muscle using electromyography. They found that even if participants tried to avoid any facial reaction, they still unconsciously reacted to the smiling face they were exposed to. One of the earliest to take a closer look at the effects of feedback were Levenson, Ekman, and Friesen (1990). They told participants of their experiment to put on certain facial expressions associated with a corresponding emotion. While participants put on the told expressions, Levenson and colleagues measured certain bodily reactions, including somatic activity. With this, they could show that displaying a specific facial expression influenced the affective state, finding evidence for the facial feedback theory. Furthermore, putting on certain expressions led to specific patterns of bodily reactions. The results were later supported by a study conducted by Ekman and Davidson (1993), who investigated spontaneous and voluntary smiling of participants while measuring their brain activity.

Current state of research on mimicry and feedback

However, despite the vast amount of research conducted concerning mimicry and feedback, the underlying processes remained not fully understood (Prochazkova & Kret, 2017). With the availability of new research methods such as fMRI, further attention was given to emotional contagion, especially to mimicry and feedback. Carr, Iacoboni, Dubeau, Mazziotta, and Lenzi (2003) found that showing facial expressions to participants led to activity in brain areas, such as the insula, the superior temporal sulcus, the amygdala, and the premotor cortex. Those areas were active also during imitation and observation of displayed emotions by other people. From this, they concluded that a first step in perceiving others' emotions is to represent them mentally. In emotion research, and especially concerning the process of mentally representing perceived emotions, decisive results were presented by Rizzolatti, Fogassi, and Gallese (2001). With the so-called "mirror neurons," they found brain cells in the brain of macaque monkeys, which are active both when a monkey carries out an activity and in situations in which it only observes it in others. These results can guide the way in understanding compassion for emotions. Path-breaking results stem from Anders, Heinzle, Weiskopf, Ethofer, and Haynes (2011). They placed couples in an fMRI study and instructed one participant to be the sender and one to be the receiver of certain emotions. They found that the received emotions' information is processed in similar brain regions that are used to produce the emotions. This representation is in scientific research referred to as "neural resonance" (Prochazkova & Kret, 2017). Anders et al. (2011) also showed a temporal delay between the neural activity in the sender's brain and the activity in the receiver's brain. Nevertheless, this delay decreased over the course of time, indicating what they called "tuning in." All those findings suggest that perceiving others' emotions is based on a somatosensory representation in the observer's brain, as previously suggested by Carr et al. (2003).

Emotional contagion through digital channels

New technological research methods were ultimately able to help by understanding the contagion process of emotions between people. However, new types of interpersonal communication, such as e-mail, raised the question among scholars as to which extent those computer-mediated communication forms limit or even inhibit emotional contagion (e.g., Lohmann et al., 2017; Neumann & Strack, 2000). In the further course, the question was raised if the contagion process follows the process known from face-to-face interactions. As was described above, emoticons and emojis are essential in everyday communication (Felbo et al., 2017). With their proceeding distribution, scientists were motivated to investigate their potential effect on emotional contagion. Different research approaches that will be discussed in more detail in the further course delivered important findings on emotional contagion through digital channels and, with this, important hints for the conceptualization of this thesis.

Early research on whether emotional contagion is possible via digital or computer-mediated channels was conducted by Neumann and Strack (2000). They exposed subjects to read out speech patterns with either cheerful or sad tones. After listening to those speech parts, they measured participants' affective state and found significantly more positive affect being reported if assigned to the positive condition. In contrast, participants reported significantly sadder affect if they were assigned to the sad condition. During the experiment, subjects could not see the person who read the text. They were connected digitally via a microphone and a headset. Hancock, Gee, Ciaccio, and Lin (2008) focused on negative affect being transmitted in the further course. They let participants interact via a text-based chat program. As they expected, participants of their negative manipulation group recognized their interaction partner's negative affect and reported for themselves a more negative affect after the interaction. One of the first studies to examine the impact of emojis on the process of emotional contagion was the study conducted by Lohmann et al. (2017). In their web-based experiment, subjects were exposed to the -emoji and, in the further, course asked about their affective state via self-report scales. The effect of emojis on the reported affective state was recently reinvestigated by Smith and Rose (2020). They conducted multiple subsequent studies, during which they presented participants the -emoji and measured their reported affect and their facial reactions. The authors found significantly more positive affect being reported and more facial muscle activity indicating mimicry when being exposed to the emoji compared to a control situation during which no emoji was shown. Taken together, the extant research provides strong evidence that emotional contagion is possible through computer-mediated channels, particularly with the use of emojis. However, it should be noted that both studies presented above were conducted in a human-to-human context. Therefore, the results may serve as a point of orientation for the further course of this thesis. However, it is unclear whether and to what extent these can be transferred to the context under consideration here.

Emotional contagion's relevance for service encounters

Various research approaches have been able to show that in evaluation situations, people do not base their decisions exclusively on rational considerations but that emotional aspects are likewise significantly involved. This also applies to the evaluation in the context of customer satisfaction (Martin, O'Neill, Hubbard, & Palmer, 2008). An explanatory approach can be undertaken based on the so-called affect-as-information theory (Clore et al., 2001). Clore et al. (2001, p. 124) conclude for their approach that "[…] emotional feelings serve as affective feedback that guides judgment, decision-making, and information processing." Schwarz and Clore (1983) found evidence for their theory by putting subjects in either a positive or a negative affective state before letting them report their general life satisfaction. In the same article, they report a second study that unveiled that people whose affect was influenced either by good or bad weather (Cunningham, 1979) reported a higher degree of life satisfaction on sunny days and a lower degree on rainy days. They furthermore concluded that affect in complex decision situations serves as heuristic information. One marketing-related field of research that heavily investigated emotional influences is research on brand management. Deppe et al. (2005), for example, showed that emotions induced by the presence of a favorite brand significantly influenced decision-making processes in buying situations.

As presented above, from a psychological point of view, emotional contagion has a long history in scientific research and its importance is undisputed. Starting in the early 2000s, the topic increasingly attracted the interest of scholars conducting service research, and they began to investigate the phenomenon in terms of its relevance for service encounters (e.g., Pugh, 2001). Around the same time, researchers began to conduct research on the so-called emotional labor as described in Section 3.2.2 (Hennig-Thurau, Groth, Paul, & Gremler, 2006). From these initiated research approaches, the two main research streams developed, which are of essential importance for service encounters. The main reason both streams attracted scholars' attention was that delivering satisfying and high-quality service is crucial for companies (Bitner, Brown, & Meuter, 2000; Heskett, Sasser, & Schlesinger, 1997; Zeithaml, Berry, & Parasuraman, 1996). In this regard, satisfaction is considered to be attributable to a specific service experience (Oliver, 1981). Scholars widely agree that satisfaction with a service encounter is a central determinant for other relevant outcomes, such as customers' behavioral intentions. For example, Athanassopoulos, Gounaris, and Stathakopoulos (2001) found support for customer satisfaction significantly influencing behavioral intentions such as, for example, intentions to return (loyalty intentions).

Earlier research found customers' emotions significantly affecting their service satisfaction (Oliver, 1993; Smith & Bolton, 2002; Westbrook & Oliver, 1991). Therefore, scholars started to investigate how far employees' affective delivery may play a role in this relationship. One of the first scholars to investigate emotional contagion in service encounters was Pugh (2001). He observed service encounters and was able to show that positive emotions displayed by service employees did not only affect customers' positive affect, but also positively impacted customers' perceptions of the service. To be precise, Pugh (2001) initiated a field study in different bank branches and investigated how the expressions of positive emotions by employees affected customers' reported affective state and their service evaluation. He was able to link displayed emotions by employees to customers' affect and customers' evaluation of the service transaction. In this way, he could show that emotional contagion does also occur in business contexts and may lead to beneficial service outcomes. Furthermore, vital implications could be drawn from his research, as he revealed the importance of displayed emotions by employees in terms of their relevance for the evaluation of the service transaction. Subsequent investigations later supported and complemented Pugh's findings (e.g., Barger & Grandey, 2006; Tsai & Huang, 2002).

Around the same time, Tsai and Huang (2002), for example, applied a similar procedure by observing service encounters in shoe stores. They found further evidence for an existing relationship between positive displayed emotions by employees and interacting customers' positive affect. In contrast to Pugh (2001), however, they linked emotional contagion not to evaluating aspects, such as encounter satisfaction, but to customers' behavioral intentions. They found that those are positively influenced by customer positive affect as well.

Barger and Grandey (2006) later affirmed the positive link between customer post-encounter affect and encounter satisfaction. What they failed to approve was a contagion process from the employee to the customer. Even though they found a correlation between employee smiling and customer smiling (i.e., mimicry), they could not link employee smiling to customers' post-encounter affective state. They suggested that this may be due to the length of their service encounter that may have been not long enough to let a certain degree of intimacy develop. What they could find, however, was that the degree of smiling by an employee predicted the degree of smiling by an interacting customer (mimicry). They could also link customers' smiling to customers' post-encounter affective state (feedback) and subsequently to encounter satisfaction. These results are relevant insofar as they mean for the later discussed empirical part that the service encounter designed in this course must meet certain requirements that will be addressed in more detail later.

While Pugh (2001) and Tsai and Huang (2002) entirely focused on subconscious processes, Hennig-Thurau et al. (2006) were the first to include more conscious aspects. Their investigation was heavily based on research on emotional labor. In their experiment, they found that it is primarily the authenticity of the displayed emotions that possesses the ability to influence customers' affective state and subsequently influence customers' satisfaction with the encounter. However, the extent of the employee's smiling was not linked to customers' affective state but rather to customer-employee rapport, which instead indirectly influenced customer satisfaction.

Most of the research on emotional contagion focused on positive and unique encounters. However, Dallimore, Sparks, and Butcher (2007) concluded that catching emotions may similarly happen from customers to employees, especially negative ones during situations of a complaint. Du et al. (2011) shifted their attention to what they called "multiple emotional contagions." They could show that multiple contagions may happen consecutively by first exposing customers to a negative scenario before exposing them to a situation where positive emotions were displayed. Their results showed that even in a situation of negative affect of a customer, the employee may trigger positive emotional contagion, which in turn significantly decreased customers' reported negative affect. The issue of emotional contagion in service encounters with potentially negative emotions being present will be taken up again and addressed separately in *Study 6*.

3.2.2 Perceived Authenticity of Displayed Emotions

One aspect that has to be taken into consideration regarding emotional contagion and its positive impacts on encounter satisfaction and other relevant outcomes is the authenticity of the displayed emotions (Grandey et al., 2005; Hennig-Thurau et al., 2006). However, reality proves it to be rather complicated to define what can be understood when talking about authentic emotions. Many authors have tried to define authenticity (e.g., Barrett-Lennard, 1998; Mesmer-Magnus, DeChurch, & Wax, 2012; Wood, Linley, Maltby, Baliousis, & Joseph, 2008). Some authors attempted to specify it via genuineness (e.g., Ashforth & Humphrey, 1993; Goldman & Papson, 1996) others approached it as the opposite of being artificial (e.g., Featherman, Valacich, & Wells, 2006) or inauthentic (Rogers, 1961). This multitude of different approaches shows that a definition of the term is not an easy undertaking. What is commonly termed as emotional labor may serve as a possibility to approach the question of what can be understood when talking about authentic emotions from a receiver's perspective. After a short introduction to the topic emotional labor, a development of a conceptual understanding of authenticity will be developed in the further course.

Many companies expect their employees to display adequate emotions that generally are of positive valence (Groth, Hennig-Thurau, & Walsh, 2009; Yoo & Arnold, 2016). For employees, this often means that they have to manage their emotional state. This "[…] management of feeling to create a publicly observable facial and bodily display […]" (Hochschild, 1985, p. 7) is, strictly speaking, referred to as "emotional labor." Hochschild (1985) was the first to introduce the term as well as synonyms, such as "emotion work" or "emotion management," when dealing with this topic in a private context. In a broader sense, emotional labor is defined as the "[…] act of displaying the appropriate emotion (i.e., conforming with a display rule) […]" (Ashforth & Humphrey, 1993, p. 90). What distinguishes the latter from Hochschild's definition is its consideration that an employee may feel the emotions required without actually having to manage them (Ashforth & Humphrey, 1993). Their definition thus just focuses on the result, the display of the desired emotions, itself. Usually, employees are expected to show positive emotions at work (Grandey, 2003). If they do feel positive and do not need to actively alter their expressed emotions, according to Hochschild (1985), this would not be considered emotional labor.

However, if the genuinely felt emotions are not in line with what is expected to be displayed, also called emotional dissonance (Johnson & Spector, 2007), employees may use two different strategies to regulate their emotions. Following the emotion regulation theory proposed by Gross (1998), affect can be modulated at two possible points, after the emotion has emerged (response-focused regulation strategy) and before its generation (antecedent-focused regulation strategy). Research on emotional labor came to the conclusion that employees may engage either in surface or in deep acting to ultimately display the appropriate emotions (e.g., Ashforth & Humphrey, 1993; Grandey, 2000; Mesmer-Magnus et al., 2012). Surface acting in this context is considered to be expressing the expected emotion but without actually trying to change the inner state (Ashforth & Humphrey, 1993). As the expression is a successor of the inner feelings that remain unaltered, surface acting is defined as a response-focused regulation strategy (Mesmer-Magnus et al., 2012). Deep acting is different from surface acting because it focuses on the antecedents of the displayed emotions, inner feelings (Mesmer-Magnus et al., 2012). An employee who is engaged in deep acting is modifying "[…] feelings to match the required displays" (Grandey, 2003, p. 87). This means that employees express expected emotions by actually trying to feel the corresponding emotion (Hennig-Thurau et al., 2006). Therefore, deep acting can be considered as an antecedent-focused regulation approach (Grandey, 2000).

It is necessary to mention that both strategies are some kind of "faking" the displayed emotions as surface as well as deep acting means that either only the expression or the expression, as well as the inner state, are modified to meet the expectations and thus do not equal what was genuinely going to be displayed by the employee (Glomb & Tews, 2004; Hochschild, 1985). When comparing both strategies, surface acting is regularly considered as being "faking in bad faith," while deep acting is referred to as "faking in good faith" (Rafaeli & Sutton, 1987, p. 32). However, with the modification of the inner affective state, the risk of a mismatch between what is felt and what is displayed is reduced (Grandey, 2003). The distinction between an authentic and inauthentic emotion that is being displayed is primarily based on subtle facial cues (Ekman, Friesen, & O'Sullivan, 1988). Thus, some scientists consider deep acting as displaying authentic emotions (e.g., Hennig-Thurau et al., 2006; Medler-Liraz, 2016). However, Ekman and Friesen (1982) conclude that trained people may be well able to display false smiles that are indistinguishable from actually felt ones. Those explanations show that a mismatch between what is displayed and what is felt (surface acting) may result in perceived inauthenticity. This risk may be reduced if there is (perceived) congruency between the inner affective state and the displayed emotions (deep acting). But in the end, also training facial expressions appears to be a possible strategy. Thus, one can conclude that it is primarily depending on the receiver who needs to feel that his/her counterpart is actually feeling the emotions he/she is displaying. To get to the heart of what can be understood of authentic emotions, based on the explanations above, the thesis in the further course builds on Mesmer-Magnus et al. (2012, p. 9), who claim: "Congruent emotional states are psychological conditions wherein an individual's authentic felt emotions are consistent with his/her expressed emotions." However, this statement shall be edited to define authentic emotions as existent *if an individual's expressed emotions are at least perceived to be consistent with the felt emotions.* Thus, the thesis takes a rather receiver-oriented view. This approach is in line with an early assessment by Rogers (1961), who, aside from any regulation strategies, approached authenticity via inauthenticity, highlighting that it is the perceived incongruence between the inner (the true) self and what is displayed that can cause inauthenticity. This approach once more underlines the importance of the receiver's perception of congruity that determines authenticity.

3.2.3 Empathy

Up to this point, this thesis primarily focused on unconscious emotional contagion between interacting people. While for a long time scientists were not able to draw the line at what is seen as the unconscious transmitting of emotions (i.e., emotional contagion) and the more conscious process of empathy, today scholars make a clear distinction between the two concepts (Decety & Lamm, 2006; Hatfield et al., 1994; Preston & De Waal, 2002). This distinction is based on the knowledge that both concepts are related, but fundamental differences exist between them. However, the extant research does not show a clear distinction in terms of the use of terms between the different processes. For example, sometimes, the relationship between emotional contagion and empathy is also described as primitive empathy and cognitive empathy (e.g., Prochazkova & Kret, 2017). This view is in line with arguments by Decety and Lamm (2006), who classify emotional contagion as a primitive form of empathy. As stated earlier in the context of the explanations on emotional contagion, in the further course, the term emotional contagion will be used when referring to the unconscious process (i.e., the unconscious transfer of emotions). When the focus is on the conscious counterpart, the term empathy will be used.

However, before empathy is considered in detail in the further course, a terminological delimitation must first be made. When looking at the scientific literature on empathy, one quickly notices that there is in many cases no clear distinction between the terms empathy and sympathy. This fact has already been pointed out by literature published some time ago (e.g., Wispé, 1986). Nevertheless, this problem is still evident today. As a first step, sympathy must be explicitly delimited from the investigated empathy, even though there are synonymous conceptual relationships in the literature (e.g., Rosenthal-von der Pütten et al., 2013; Rosenthal-von der Pütten et al., 2014). This distinction is necessary for two reasons: sympathy, while at first sight being similar to empathy, is related to a person's reaction toward negative emotions or another person suffering (Decety & Chaminade, 2003). A goal-oriented approach to differentiating between the two constructs of empathy and sympathy comes from Wispé (1986). The approach claims that, while feeling sympathy, there is a lost form of self-differentiation and communion is built through understanding the emotional state with the aim of improving the "well-being" of the other person. Wispé (1986) further emphasizes that sympathy is de facto more than the perception of others' negative emotional states. In another work, Wispé (1991, p. 68) suggests as a definition of sympathy by stating that "[...] sympathy has two parts: first, a heightened awareness of the feelings of the other person, and, second, an urge to take whatever actions are necessary to alleviate the other person's plight." This view is supported by Davis (1994), who stresses the desire that is inflicted by sympathy to mitigate the observed suffering. The need for a distinction between sympathy and empathy additionally results from a different use of terms in the German language compared to the English-language scientific literature. While in the German language, sympathy often emphasizes a person's positive affection toward another person, scientific literature uses the Greek and Latin word origins as a basis (Wispé, 1986). For example, the Greek word *sympátheia*, composed of the components *sýn* (with) and *páthos* (suffer), means de facto "to suffer with someone."

Along with the long history of empathy in scientific research, there is an equally long debate on defining it (e.g., Clark, 2007; Davis, 1994). On the one hand, some authors argue in favor of an "affective empathy" and thus stress the affective component (e.g., Eisenberg & Miller, 1987; Mehrabian & Epstein, 1972). The approach of affective empathy is sometimes referred to as "emotional empathy" (e.g., Mehrabian & Epstein, 1972). Gladstein (1983, p. 468), for example, defined it as "[...] responding with the same emotion to another person's emotion." The affective approach comes close to the initially proposed German *Einfühlung* suggested by Lipps (1907) and it shows substantial overlap with the previously discussed concept of emotional contagion as it contains an adoption of other's emotions.

On the other hand, scientific research always comes back to the aspect of cognition in the course of empathy and thus to the authors who speak of "cognitive empathy" and hence stress the ability to understand affective states of other people (Baron-Cohen & Wheelwright, 2004). The fundamental differences between affective and cognitive components of empathy were introduced into the discussion by Mead (1934). In the context of cognitive empathy, the supporters of this approach emphasize that the focus is on understanding the emotions of others and, with this, required cognitive capacities to do so (Feshbach & Roe, 1968). For an extended period, the so-called theory of mind was also discussed in the course of cognitive empathy (e.g., Baron-Cohen & Wheelwright, 2004; Roeyers, Buysse, Ponnet, & Pichal, 2001). Gallagher and Frith (2003, p. 77) define theory of mind as having the "[...] ability to explain and predict the behaviour [sic] of ourselves and others by attributing to them independent mental states, such as beliefs, desires, emotions or intentions." The positions discussed on the differentiation between affective empathy and cognitive empathy are, in some cases, more or less strict. For the further course, however, theory of mind is not further consulted as regarding emotions, the concept of empathy provides a narrower approach than does theory of mind. Based on the cognitive effort expended, other authors emphasize a conscious empathy process (Batson et al., 1991; Davis, 1994). While emotional contagion focuses on the (unconscious) transmission of emotions, it views empathy as a process of putting oneself in the role of an interacting person. This aspect was introduced comparatively early by Rogers (1959) when he stated:

> The state of empathy, or being empathic, is to perceive the internal frame of reference of another with accuracy, and with the emotional components and meanings which pertain thereto, as if one were the other person, but without ever losing the "as if" condition. Thus, it means to sense the hurt or the pleasure of another as he senses it, and to perceive the causes thereof as he perceives them, but without ever losing the recognition that it is as if I were hurt or pleased and so forth. If this "as if" quality is lost, then the state is one of identification. (p. 210-211)

Wispé (1986) describes empathy as a person's attempt to understand another person's experiences, whether positive or negative. Furthermore, she attributes a certain amount of effort and thus consciousness to empathy. This aspect of consciousness and the associated cognition set empathy apart from the previously discussed emotional contagion. Concerning the relationship of these constructs to each other, Preston and De Waal (2002) conclude that all the investigated aspects of emotional contagion, empathy, and sympathy do, however, share similarities and may thus not be investigated entirely separately. This aspect is mainly accepted in scientific research today, such that there is an agreement that empathy includes an affective response to the emotions of another person (Davis, 1983; Decety & Jackson, 2004). In addition, the authors note that empathy also includes the cognitive component outlined above, which allows understanding and mentally representing affective states of other persons. However, it is important to stress that, while the distinction between the two affective states blurs during emotional contagion, with empathy, there remains a separation between the own feelings and the feelings of the other person (Decety & Jackson, 2004; Rogers, 1959). That is why today usually there is no strict separation into cognitive empathy and affective empathy anymore, but empathy is seen instead as a multidimensional construct consisting of both dimensions (Baron-Cohen & Wheelwright, 2004; Rogers, Dziobek, Hassenstab, Wolf, & Convit, 2007). Now that the basic characteristics of the underlying understanding of empathy have been outlined, another further delimitation shall be made, relevant for the further course of this thesis, before the current state of research is discussed. Looking at the existing research on empathy, it becomes apparent that research approaches this complex of topics in two ways. It follows that with the term empathy, both the "predictive" (e.g., Dymond, Hughes, & Raabe, 1952; Rogers et al., 2007) and the "situational" perspective (e.g., Borke, 1971; Jackson, Brunet, Meltzoff, & Decety, 2006; Ruby & Decety, 2004) are addressed. While the predictive perspective is based on a more general understanding of empathy as the ability of the observed individual to put himself/herself in other people's shoes, the situational perspective usually looks at a specific situation. It focuses on the question of whether an individual in the situation can empathize with another person. This separation of the two perspectives is relevant insofar as this thesis will specifically adopt the second perspective (i.e., emphasizing with the affective state of a chatbot) in the further course.

Just like research on emotional contagion, research on empathy has also developed in recent years as new research methods have broadened the understanding of the underlying processes. Groundbreaking insights about empathy similarly come from the field of neurosciences through methods such as fMRI or positron emission tomography. As a clear distinction between emotional contagion and empathy only developed over the years, many of the presented findings concerning emotional contagion also added to the literature on empathy. However, while some authors speculated that contagion (i.e., feeling what the other person is feeling) is a necessary prerequisite for empathy (e.g., Gallese, Keysers, & Rizzolatti, 2004; Keysers & Gazzola, 2006), new research approaches supported what was previously already referred to as the distinction between self and other. For example, besides their decisive role in emotional contagion, mirror neurons also seem to be involved in the process of empathy (Anders et al., 2011). This indicates that empathy builds on the same brain areas being activated while observing or imitating emotions and feeling emotions. However, there seem to be fundamental differences, especially concerning the strength of the activation in certain areas. For example, Ruby and Decety (2004) conducted a positron emission tomography study during which participants of the study were shown texts telling situations taken from everyday life and that were determined to induce specific emotions. As a second step, they were asked to imagine their mother's perspective or their own perspective if they were in the shown situation. With this procedure, the authors could show that multiple of the investigated brain areas were similarly activated across both scenarios. There were, however, areas of the brain that showed differences concerning their activation. Those areas were the left temporal pole as well as the right postcentral gyrus. The authors concluded that those results support the idea that empathizing with others is primarily based on representing the emotions in their own brains. Nevertheless, there remains a clear distinction between the empathized emotions and the own affective state (Decety & Jackson, 2004; Rogers, 1959). The results described above were supported by Jackson et al. (2006). They designed their study to inflict imaginary pain by showing pictures of situations where pain was inflicted. Participants were asked to imagine being in a certain situation or another person. The authors found an activation of the neural network that is responsible for processing pain in both cases. However, the activation was activated more strongly when participants imagined themselves in the situation. From both studies, Decety and Jackson (2006) concluded that the following areas in the brain must be responsible for differentiating whether you feel emotions yourself or put yourself in the position of another person: the posterior cingulate, the precuneus, and the right temporo-parietal junction. These results have crucial implications for both scientific research and the rest of this thesis: the distinction between the perspective of oneself and the one of others in terms of felt emotions is a critical aspect of empathy, as it does aim at a clear distinction between the involved people (Decety & Jackson, 2006). This point also distinguishes empathy (i.e., emphasizing with the affective state of a chatbot) from (primitive) emotional contagion presented above, where, in the end, there is no longer any difference between the affective state of the sender and that of the receiver.

4 Overview of Current Literature

The present thesis combines several research streams, such as research on emotional contagion and empathy in the context of artificial entities as well as computer-mediated communication in a human-to-human context. To provide a concise overview of the current literature, which will form the basis for the empirical studies, the following section will outline the current state of research of the research streams used and present the underlying limitations of those results. Figure 4.1 will finally graphically summarize the current state of research and the outlined research gaps. It is becoming clear that the research gaps are, in a sense, to be viewed separately. While *Study 1* to *Study 4* focus strongly on basic questions of whether a chatbot is able to trigger emotional contagion and empathy, *Study 5* and *Study 6* strive to extend previous results and focus on the exploration of boundary conditions.

In service marketing, a service encounter is the central situation that decides on a positive or negative further course of the customer relationship (Oliver, 1981). In this context, satisfaction with the service provided is the central success factor from the customer's point of view (e.g., Athanassopoulos et al., 2001; Heskett et al., 1997). Regarding encounter satisfaction, it is undisputed that the affect-as-information theory (Clore et al., 2001) plays a role in these situations and that customer's affective state is used for the evaluation. Based on these considerations, emotional contagion has found its way into service research. The extant research has shown that the affective delivery of an employee can influence the affective state of the customer, which in turn has a positive influence on the evaluation of the service encounter (e.g., Pugh, 2001). With the increasing spread of chatbots and the growing replacement of frontline employees with chatbots, the question arises to what extent a chatbot, which communicates exclusively via text-based communication channels and is characterized by its disembodied nature, can change the affective state of a customer. The increasing use of

K. Prinz, *The Smiling Chatbot*,
https://doi.org/10.1007/978-3-658-40028-6_4

chatbots and the knowledge of the extant research that, in a human-to-human context, service encounters are also evaluated on the basis of emotional components finally raises the question of whether these mechanisms are valid in chatbot-handled encounters. With the spread of AI technologies, initial research has already addressed whether the contagion of emotions from artificial entities to humans is possible (e.g., Tielman et al., 2014; Tsai et al., 2012). However, the validity of the results is limited in the present context because the results of Tsai et al. (2012) are based on a graphical representation of the artificial agent. Furthermore, the results of Tielman et al. (2014) are based on a sample consisting of children and thus did not have a service focus.

In addition to emotional contagion, research on this topic repeatedly considers empathy, which is generally regarded as the cognitive counterpart to emotional contagion (Hatfield et al., 1994). As was pointed out above (e.g., Decety & Jackson, 2006), the main difference is that empathy toward the affective state of a counterpart leaves a clear distinction between the two affective states. The intense scientific debate about empathy has not gone unnoticed by research on AI, which has devoted itself to the question of how far artificial entities are capable of triggering empathic reactions. In this respect, research on robotics has been leading the way. In this context, Rosenthal-von der Pütten et al. (2014) found that similar patterns occurred in test persons' brains concerning the attributed empathy toward either a human or a robot. However, the research results mentioned above have substantial limitations for two reasons. First, they did not occur in a service context. Second, the focus was on negative emotions, as, for example, pain was inflicted to the robot. As already outlined above, the approaches lead primarily in the direction of sympathy and therefore have little relevance for service research. Furthermore, robots are generally ascribed a higher potential to trigger anthropomorphic behavior toward them due to their physical appearance compared to, for example, chatbots (Duffy, 2003). Therefore, it cannot be assumed whether and to what extent the effects described above are applicable in the context of chatbots. This gap is relevant for service marketing, since very few research approaches have shown that empathy toward an employee impacts client-employee relationships (Wieseke et al., 2012). In this respect, the question arises to what extent these effects also occur during an interaction with a chatbot.

The primary basis for the present thesis are the results of human-to-human research in the field of computer-mediated communication. Since, in general, more and more communication and service encounters are taking place via digital channels, scientific research has increasingly focused on the question of the representation of emotions via digital channels in recent years (e.g., Lohmann et al., 2017; Smith & Rose, 2020). The question of emotional contagion has not been ignored by this research stream either, so that it has been investigated whether and to what extent contagion via digital channels or computer-mediated communication is possible. In a first step, Lohmann et al. (2017) have shown that displaying positive emotions using an emoji and showing it to an exclusively female sample has led to significantly more positive affect being reported in a private context. Smith and Rose (2020) further investigated the use of emojis in the context of a service encounter. Not only did they find that emotional contagion is possible with the help of emojis in computer-mediated communication, but they were also able to show that the presentation of emojis causes facial mimicry. However, the results in this research stream underlie the main limitation that both research approaches are limited to human-to-human communication. For this reason, it is ultimately not possible to assess whether and to what extent the processes known from human-to-human communication can be transferred to human-to-chatbot interactions. There is also the question of how the processes behave in the context of service encounters. A combination of the shown research streams results in the research gaps for *Study 1* to *Study 4* outlined in Figure 4.1.

The question that arises in the further course concerns the exploration of boundary conditions of the effects resulting from the affective delivery of a chatbot (i.e., the exploration of situations in which the found affective responses by customer to a chatbot's displayed emotions may not occur). Research on anthropomorphism has shown that the tendency to ascribe human-like attributes to artificial entities can be positively influenced by perceived human likeness (Lin et al., 2021). This means that the more human-like an agent is perceived (e.g., due to its appearance), the more people tend to show anthropomorphic behavior toward those agents. For the perception of human likeness, so-called human-like cues are important (Blut et al., 2021). However, what has not yet been researched in this context is the interaction between human-like behavior and the graphical representation of a chatbot. This gap, therefore, builds the basis for the research gap to be filled by *Study 5* by exploring boundary conditions (Figure 4.1).

Furthermore, research on chatbots and their use in service encounters has shown their application being not exclusively associated with desirable outcomes. In a sales context, Luo et al. (2019) showed that customers were less willing to buy when they interacted with a chatbot. Also outside of routine activities, chatbots do not seem to be accepted by customers without reservations. A study by Mozafari et al. (2020) showed that in a critical encounter (i.e., a monetary-related mistake happened beforehand) customers trusted a chatbot less in the following service encounter. In addition, research on the use of emojis to substitute facially expressed emotions has shown that the use of emojis may backfire. For example, Thies et al. (2017) report the use to be perceived as juvenile. Duijst (2017) reports in a time-sensitive context that customers have described the use of emojis as strange. Thus, the two aspects opened up the second field to explore boundary conditions (*Study 6*). Specifically, the question remains unanswered whether the display of positive emotions by a chatbot leads to beneficial outcomes from the perspective of a service provider in critical service recovery encounters. At the same time, the question arises whether the course of the recovery (i.e., successful or unsuccessful) has an influence on the cognitive process of empathy toward the chatbot (Figure 4.1).

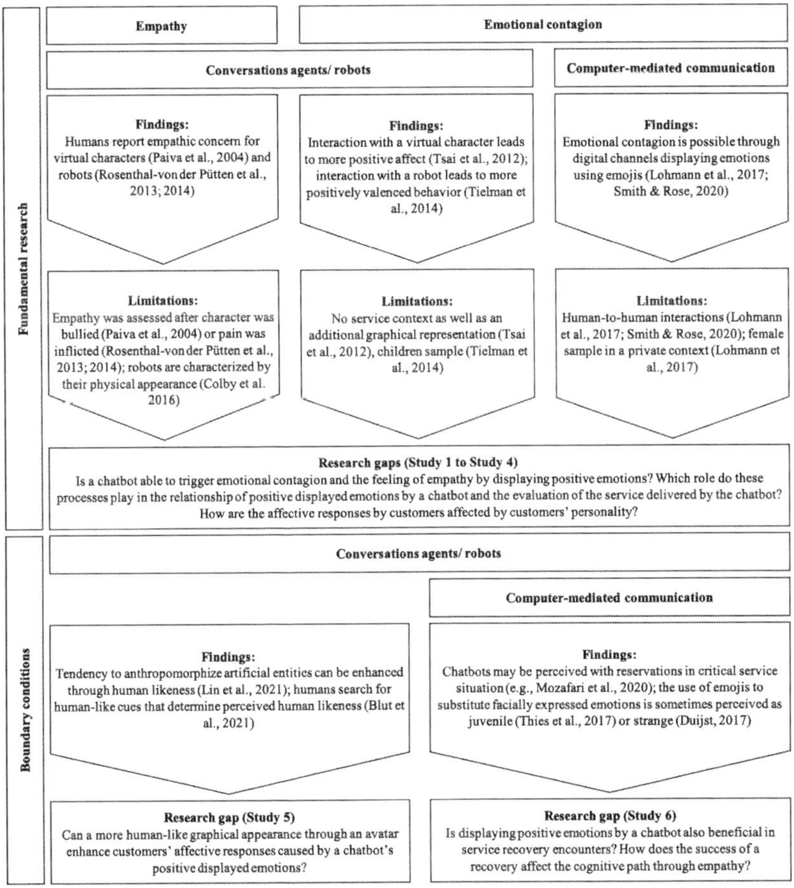

Figure 4.1 Overview of Current Literature and Derived Research Gaps

5 Conceptual Developments and Empirical Investigations

5.1 Overview

The empirical part of this thesis comprises six consecutive studies, which will be presented in the further course. *Study 1* was designed to undertake a general approach to the process of emotional contagion in a human-to-chatbot interaction. With this, the goal of *Study 1* was to test the proposition of a facial reaction of subjects to positive emotions displayed by a chatbot. For this purpose, a laboratory experiment was set up. A Wizard of Oz method, where a human operator secretly acts as chatbot, was chosen to simulate the chatbot and facial recognition was selected to measure changes in customers' facial valence. In this way, this study was the first to provide evidence for a chatbot's ability to trigger emotional contagion by displaying positive emotions. *Study 2* proposed that positive emotional expression by a chatbot influences encounter satisfaction through customer positive affect (emotional contagion). To test this proposition, video stimuli were developed and used in a web-based experiment. A booking process taken from the hospitality industry served as the basis for designing the used stimuli. Although subjects during *Study 1* and *Study 2* were led to believe that they were interacting with a chatbot, both studies relied on chatbot answers that were actually written by a human (e.g., the wizard acting as chatbot during *Study 1*). *Study 3* was supposed to overcome this limitation by using a real chatbot programmed to have subjects book a fictive hotel room during a web-based experiment. While *Study 2* and *Study 3* were limited to (primitive) emotional contagion, *Study 4* extended the proposed research model by adding empathy as a second and more conscious mediating path leading from positive displayed emotions to encounter satisfaction. Since it was also assumed in this course that the affective reactions to the positive emotions of the chatbot would not apply equally to all customers,

K. Prinz, *The Smiling Chatbot*,
https://doi.org/10.1007/978-3-658-40028-6_5

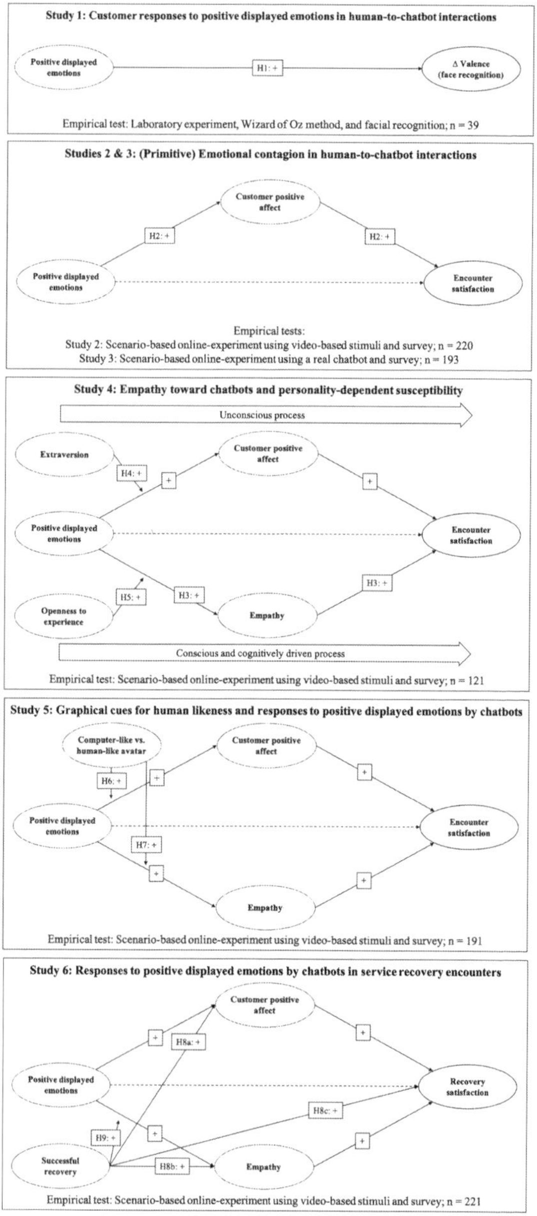

Figure 5.1 Overview of Conceptual Developments and Studies

extraversion and openness to experience as personality traits were investigated regarding their moderating effects. The results of the first four studies provided clear support for the investigated effects of positive emotional expressions of a chatbot. Building on these results, the following two studies were designed to explore boundary conditions. *Study 5* assumed that the reactions of the customers to the emotions of the chatbot could be increased by additionally displaying the chatbot graphically with a human-like avatar. To test this assumption, the chatbot was additionally represented with a human-like avatar. Additionally, counteracting effects were tested by having a second experimental group interact with a chatbot represented by a computer-like avatar, which highlighted the technical nature of the chatbot. This study yielded interesting results regarding the interplay and benefit of human-like cues. It was the subject of *Study 6* to drop the scenario of a room reservation, which had characterized a routine activity so far, and explore the validity of the found results in critical service encounters. For this purpose, a recovery situation was selected that was preceded by an error that was not the fault of the chatbot itself. The study tested a successful recovery (i.e., the chatbot can directly solve the problem that has arisen) and an unsuccessful recovery (i.e., the chatbot has to refer the customer to the customer service available by telephone). For the study, the expectation was that the successful course of a recovery situation would moderate the cognitive path through empathy. Figure 5.1 provides an overview of the six conducted studies.

5.2 Study 1: Facial Reactions to Displayed Emotions (Laboratory Experiment)[1]

5.2.1 Conceptual Development

Research has already addressed the fundamental question about whether and to what extent humans exhibit response patterns toward AI technologies, such as chatbots, that are known from human-to-human interactions. The basis for the expectation that those patterns are like to expect, stems from the so-called computers are social actors (CASA) paradigm (Reeves & Nass, 1996). The CASA paradigm dates from the course of the 1990s that brought significant new insights about media research and, particularly, people's psychological reactions. The core statement of this paradigm is that people react in interactions with computer-based agents in a way that is similar to how they would interact with humans.

[1] Parts of *Study 1* have been presented at the 7th Rostock Conference on Service Research.

The human brain, originating from a time when only humans showed social behavioral patterns, is the reason for this reaction. This leads to even the smallest patterns, which indicate social behavior, triggering social responses in the human brain. Reeves and Nass (1996) as well as Katagiri et al. (2001) emphasize that this is an unconscious process. This application of those behavioral patterns results from a search process where users look for similarities between the own characteristics and the characteristics of the agent (Katagiri et al., 2001). These search patterns serve to control the own behavior to check whether this is adequate for the interaction. However, it should be noted that the CASA paradigm's core statement is not entirely undisputed. In their study, Shechtman and Horowitz (2003) found the knowledge of interacting with a computer program led to less effort subjects showed to establish a relationship with the other entity compared to an interaction with a human being. Thus, research around the CASA paradigm in parts shows ambiguous results. For this reason, Shechtman and Horowitz (2003) warn against applying the underlying mechanisms without reflection. This advice is supported by empirical results provided by Hill, Ford, and Farreras (2015), who showed that subjects in an interaction with a chatbot adapted their communication behaviors to communicate using simpler language compared to an interaction with other humans. Support for the expectation that the CASA paradigm's mechanisms are transferable to the relationship between humans and AI comes from research on robotics (e.g., Kim et al., 2013). The assumption of the underlying hypothesis (i.e., the CASA paradigm's validity in the present context) has already been brought into play by Heerink, Kröse, Evers, and Wielinga (2010), who assumed that, in analogy to the treatment of computers as social actors, one could expect the paradigm's applicability in the context of conversational agents. However, Heerink et al. (2010) and von der Pütten et al. (2010) emphasize that treating a conversational agent as a social entity depends primarily on its graphic representation. In the end, this would imply that the paradigm maintains its validity for embodied conversational agents but not for chatbots without any graphical representation. That there is obviously a difference in perception between humans and chatbots is demonstrated by results from Mozafari et al. (2020). They found that subjects perceive a counterpart as significantly less human if they know that the conversation partner is a chatbot. Thus, there seems to be no general guarantee that the expected processes from other research contexts are automatically transferable to interactions with chatbots.

When talking about typical human behaviors, one must inevitably consider the expression of emotion (Ekman, 2000; Ekman et al., 1980; Myers, 1989) and emotional contagion (Hatfield et al., 1992). Emotional contagion between people is a three-step process (Barsade, 2002). First, (unconsciously) the emotions perceived by the other person are mimicked. Second, this mimicry then causes a subconscious adaption of the inner emotional state of the receiver. Finally, the altered emotional state affects how one feels at a conscious level. As was presented above, all steps of the process are well documented for face-to-face interactions. This includes facial, vocal, or postural mimicry as well as the feedback process (see Section 3.2.1) that in turn leads to a change of the affective state. However, just recently, scholars entered the field of investigating those steps also in a digital and computer-mediated context (e.g., Lohmann et al., 2017; Smith & Rose, 2020). Those approaches could show that emojis possess the capabilities to trigger facial mimicry. The latest support for the expectation that emojis, used as a substitute for facially expressed emotions, may trigger emotional contagion stems from human-to-human-based research conducted by Smith and Rose (2020). They used the -emoji in a laboratory study to emotionally enrich messages and found that participants responded with significantly more self-reported positive affect when they belonged to the manipulation group compared to a control group with no emojis displayed. Most notably, however, their participants did respond with significantly more smiling after being exposed to the stimulus that contained the emoji. From this, they concluded that emojis are well able to elicit facial mimicry, which in turn impacts the affective state, which is similar to the effects reported from face-to-face interactions.

In the context of emotional contagion, a research paper by Tsai et al. (2012), builds the basis for the expectations around this phenomenon can be expected to occur from a chatbot to a human. They examined emotional contagion in the context of a virtual character. Across multiple studies, they exposed subjects to virtual characters expressing emotions. They found subjects to report more happiness if exposed to a character that displayed a happy face. Similar results, however, using a robot, were provided by Tielman et al. (2014). They found evidence that a robot can bring about affective change. They allowed children to interact with different robots that either showed affective behavior expressed through voice, body language, and gestures or neutral behavior. The authors found that children from the manipulated group who interacted with the robot that displayed affective behavior also displayed more positively valenced behavior. Both studies show that the transfer of an affective state from an artificial entity to a human being seems to be generally possible. However, in both studies, the agent had at least a graphical (Tsai et al., 2012) if not a physical representation (Tielman et al., 2014). Nevertheless, together with the results of Smith and Rose (2020), these results motivate the expectation that a chatbot even without any graphical representation can trigger emotional contagion.

Typically, employees in service transactions are expected to display positive emotions (Groth et al., 2009; Yoo & Arnold, 2016). Thus, because of the research context of this thesis, the expression of and response to positive displayed emotions by a chatbot is of particular interest. Should there be a mimicry of the emotions expressed by the chatbot, the observed facial expression of the subjects would have to be characterized by a correspondingly more positive valence. Thus, the participant's valence is considered a good indicator of any facial reactions (mimicry) to the chatbot's displayed positive emotions. Using the valence as a basis for evaluating emotional responses is further encouraged because this target value was already part of previous research to capture body-related mimicry reactions to emotive expressions of an artificial entity (e.g., Tielman et al., 2014). Taken together, this means that for *Study 1* it is assumed that the expression of positive emotions, a typical human behavior, which is even a necessity in service encounters, leads to the chatbot being perceived by customers as a social actor. This perception as a social actor, in accordance with the CASA paradigm, then causes customers to mimic the displayed positive emotions, which ultimately results in a more positive valence of the facial emotions. Therefore, the following hypothesis is proposed (Figure 5.2):

> **H1**: The display of positive emotions by a chatbot leads to more positively valenced facial reactions of customers compared to a situation when the chatbot displays no emotions.

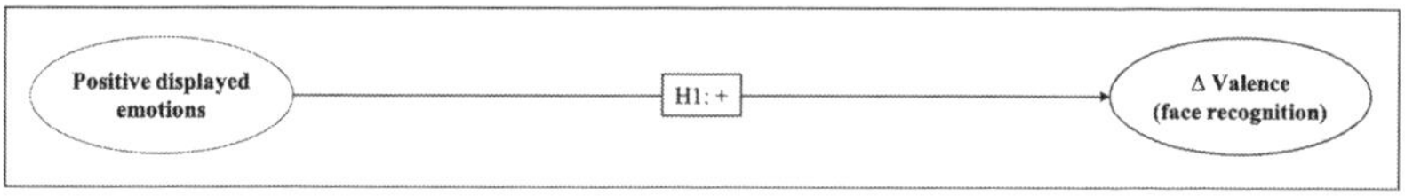

Figure 5.2 Proposed Research Model and Hypothesis, Study 1

5.2.2 Research Objectives and Setting

Study 1 was designed to fundamentally shed light on the process of emotional contagion between a chatbot and a human. For an optimal investigation of customers' reaction to the positive emotions of the chatbot, the physical presence of the subjects was necessary. Thus, for this basic examination, a laboratory experiment was set up as this enabled to observe and analyze unconsciously occurring processes during customers' interactions with the chatbot. In addition, there is typically the greater control of side effects associated with laboratory experiments (Homburg, 2018).

First, a research setting had to be identified before, as a second step, a service encounter related to this setting could be chosen. For the presented study, a hospitality industry setting was selected. Two reasons were decisive for this decision. First, the extant research on emotional contagion also favored a hospitality industry setting. For example, Grandey et al. (2005) chose a check-in encounter. In their experiment, the authors found evidence for emotional contagion happening between the interaction partners. Those findings indicate that a hospitality industry setting seems to be well-suited for research on emotional contagion. Second, delivering quality service in this industry is crucial (Grandey et al., 2005). This need leads to an apparently high willingness to adopt new technologies to improve customer service quality (Mongin, 2020). This willingness, maybe even a necessity, is further supported by increasing competition in the accommodation industries (Tsaur & Lin, 2004).

As a second step, based on the chosen setting, a booking process appeared promising for the encounter simulated during the experiment. Multiple reasons supported this assumption. (1) Considering a typical process of a private hotel stay, going through the booking process is an inevitable process stage. Thus, it is reasonable to assume that many people would be able to put themselves in the situation of the customer, as they probably have experienced such a situation before. Besides, this step is of rather short duration in the case of short trips, which made such an encounter well suited for the use in the present study. (2) Thinking, for example, about a chatbot that acts as a personal assistant, like Siri or Alexa, looking for and booking hotels via chats appeared as a realistic future field of application for chatbots. (3) As emojis are a possible substitute for facially displayed emotions in personal contact (Felbo et al., 2017), the context must be suitable for using those substitutes. In other settings that, for instance, require urgency, emojis are perceived as inappropriate (Duijst, 2017). The recently conducted research by Smith and Rose (2020) favored a scenario around an appointment reservation, further encouraging to choose a booking scenario, thus showing a high degree of commonality.

5.2.3 Experimental Design

Chatbot design

One central topic in setting up the laboratory experiment concerned the general design of the chatbot. Instead of programming a chatbot, a Wizard of Oz method

(Dahlbäck, Jönsson, & Ahrenberg, 1993) was favored. This method relies on telling subjects that they are interacting with a chatbot when, in reality, the chatbot is controlled by a human confederate. This is a common method in research on human-robot interactions (e.g., Stock & Merkle, 2018). Furthermore, different approaches to investigate interactions between humans and computers or chatbots favored this method (e.g., Bickmore & Cassell, 2005; Ho et al., 2018; Skjuve, Haugstveit, Følstad, & Brandtzaeg, 2019; Thies et al., 2017). One reason for choosing this method over applying a real chatbot was the greater extent of control over the flow of the conversation and thus the intended manipulation. Another reason for choosing the Wizard of Oz method was its well-balanced ratio of internal and external validity (Ho et al., 2018; Homburg, 2018). External validity was ensured by having a participant interact with what they perceived as a real chatbot. Internal validity was firstly ensured by giving the participants a defined task that should be completed during their interaction with the chatbot. This ensured, although the chats offered a high level of flexibility, a certain degree of comparability. This means that they were asked to book a hotel room for two people at a hotel in Munich. Besides that, internal validity was ensured because the Wizard of Oz method offered the potential to standardize the chatbot's behavior across participants. This was achieved by developing a chart to identify possible subject messages and assign corresponding chatbot answers (Table 5.1 provides an excerpt of the prepared answers). Aspects that were of particular interest were, for example, the hotel suggestions provided by the chatbot. This applied, for example, if a customer was not satisfied with the suggested hotel. For those situations, multiple options of different price categories were prepared, consisting of a hotel description and corresponding pictures. Lastly, internal validity was furthermore ensured by having only one person acting as the chatbot across all interactions.

Table 5.1 Prepared Answers for the Wizard of Oz Method, Study 1

Topic	Chatbot answer (with positive emotions)	Chatbot answer (without positive emotions)
Welcome message	Hello, my name is Lisa. I am your personal chatbot. I am glad you are here 😀. I can help you with bookings or other concerns about our hotels. What can I do for you?	Hello, my name is Lisa. I am your personal chatbot. I can help you with bookings or other concerns about our hotel. How can I help you?

(continued)

Table 5.1 (continued)

Topic	Chatbot answer (with positive emotions)	Chatbot answer (without positive emotions)
Asking customer for travel date	Beautiful city 😀. When would you like to arrive and depart?	With pleasure. When would you like to arrive and depart?
Confirming travel information	So, we now have two nights for two people for the period 11.12.–13.12. in a double room in Munich. One moment, please. I am generating a suitable offer for you 😀.	All right. We have two nights for two people for the period 11.12.–13.12. in a double room in Munich. One moment, please. I am generating a suitable offer for you.
Hotel suggestion (example)	Then, we have the Infinity Munich. It is centrally located and only 1.1 kilometers away from the Marienplatz 😀. Here are a few pictures to look at: [Image thumbnails were sent afterwards]	Then, we have the Infinity Munich. It is centrally located and only 1.1 kilometers away from the Marienplatz. Feel free to get an impression of the hotel: [Image thumbnails were sent afterwards]
Booking confirmation	Thank you for your reservation 😀. The room is booked for you with the reservation code KO119763. Please state this code at check-in.	Thank you for your reservation. The room is booked for you with the reservation code KO119763. Please state this code at check-in.
Farewell message	I am glad that I could help you. Wish you a nice stay in Munich and see you soon 😀.	I am glad that I could help you. Wish you a nice stay in Munich and see you soon.
Backup response	Unfortunately, I did not understand that. Would you like to […] instead?	Unfortunately, I did not understand that. Would you like to […] instead?

Notes: Messages have been translated from German for this table.

Display of positive emotions

The second central question concerned the behavior of the chatbot and especially the affective delivery. As part of a course held under the author's direction, a series of pre-studies was conducted that aimed at how authentic positive emotions need to be displayed through computer-mediated communication. During those pre-studies, videos were used to simulate a fictive service encounter. These videos aimed to display one chat in which no emotions were expressed and one conversation in which authentic positive emotions were shown (see Section 3.2.2 for a discussion regarding authenticity). Other than that, the chats were supposed to be identical. Due to the university environment, participants of these pre-studies were a total of 241 primarily students of a mid-sized German university. As the first step of these pre-studies, we tested different combinations of emojis and emoticons paired with a neutral tone of written text as potential substitutes for facially displayed emotions (n = 135). Among others, we tested the 😊-emoji that has been used in recent research on emotional contagion in text-based human-to-human interaction (Smith & Rose, 2020). However, in our study, pairing this emoji with a relatively neutral tone of written text resulted in significantly lower ratings of perceived authenticity of the displayed emotions compared to the exact text without any emojis. This was unsurprising, given the findings from existing research showing that individuals find it incongruous or strange when emojis are combined with a certain degree of formal communication (Duijst, 2017). Furthermore, we tested wit, which means a tendency toward informal language, alone concerning the effects of the ton of the written parts on authenticity. But, taken for itself, it performed significantly worse compared to the control chat. Consequently, as a second step of the pre-studies, we tested two different emojis (😃 and 😊) combined with a different text style (n = 106). Results indicated that the 😃-emoji paired with wit is perceived as the most authentic display of positive emotions. This equals the method used by Lohmann et al. (2017), who also conducted their research in an apparently German-speaking context. The 😃-emoji is generally associated with a "grinning face with big eyes" (Emojipedia, n.d.-b). However, the 😊-emoji is also commonly used to express smiling, which raised the question of why the latter performed worse. We assumed these results to be due to the private context the 😊-emoji is usually used in. Thus, it may have seemed inappropriate and, therefore, inauthentic when this emoji is used in a more formal business context. Based on these findings, the 😃-emoji, combined with carefully added parts of wit (i.e., informal language), was used to develop the chat behavior (see below). Based on these findings, for the first study of this thesis, the manipulation was a 2×1 between-subjects design with positive displayed emotions being the manipulated factor.

Chatbot behavior
Design principles suggested by the extant literature on chatbots and conversational agents were considered for the fictive chatbot's appearance. This concerned especially the chatbot's disclosed capabilities as well as its nature, gender, and personality. Previous research on conversational agents has shown that the gap between expectation and performance is one of the main challenges in interactions between humans and chatbots (Luger & Sellen, 2016). As customers expect that the chatbot tells the user its capabilities and which tasks it can execute to reduce this expectation-performance gap (Jain et al., 2018), the first sentence was designed to give the customer information on which tasks potentially could be performed by the chatbot. Furthermore, the chatbot's first message was designed following the findings on disclosure of chatbots in service interactions. As people expect chatbots to disclose their nature (e.g., Mozafari et al., 2020), it was explicitly revealed in the first sentence. As was discussed at an earlier stage of this thesis, people attribute gender stereotypes to chatbots and conversational agents in general (Eyssel & Hegel, 2012; Forlizzi et al., 2007; Kuchenbrandt, Häring, Eichberg, Eyssel, & André, 2014). In this regard, people's preference for chatbots whose behavior fits gender stereotypes typically related to the performed task was shown (Eyssel & Hegel, 2012; Forlizzi et al., 2007). Hence, to meet gender role expectations, the chatbot was female as recent statistics in Germany indicate a predominant portion of the employees working in customer service being female (Daum, Holtgrewe, Schörpf, & Nocker, 2018). Following findings on chatbots' personalities and people enjoying it more to communicate with a chatbot having a personality that matches its business context (Jain et al., 2018), the chatbot was designed to act like it had a real personality by telling the customer about a local event that it was about to attend.

5.2.4 Procedure

To build up the storyboard as realistic as possible, participants were invited to the experiment, telling them that the involved research group had developed a chatbot to be used in a hotel environment to improve booking processes. Before going live, this chatbot should be tested as part of a laboratory study. Upon arrival of the participants at the facility where the experiment took place, this briefing was repeated, telling them that they would chat with a chatbot and informing participants that their facial expressions would be recorded and evaluated following the experiment. Due to privacy reasons, they had to give their consent in written form besides a health declaration. No participant rejected one of the declarations, so for none of them, the experiment had to be terminated. Participants did not know about the real purpose of the experiment and that they interacted with a human chat partner instead of a chatbot. Furthermore, they received monetary compensation for their participation. They were seated in front of a 27-inch display. On this display, a webcam that recorded the participant's face was placed. A second webcam was placed upon the table facing in the direction of the monitor and recording it. The two video streams were monitored and steered using Open Broadcaster Software (OBS) Studio v24. All were recorded with 1080p and 60 frames per second. For the interaction with the computer and the chatbot, they could use a keyboard and a mouse. Shortly after each participant was seated, one of the supervisors, the one that was responsible for monitoring the technical equipment, indicated to the confederate, who was seated in another room and who acted as the chatbot, that the experiment was ready to be started. For this, the involved supervisors had a separate communication channel via a chat platform that only served to exchange information. Upon receiving the information, the confederate started typing a predefined welcome message, indicated on the subject's screen with "Chatbot is typing." From this moment on, from the perspective of the participants the chats went, besides the given task, utterly free, with only one conversation being interrupted because the participant accidentally closed the chat window. All chats ended with a reservation and the chatbot wishing the subject a pleasant stay. After the experiment, participants were still not told about the purpose of the experiment or that they had not interacted with a real chatbot. This decision was made because of the expectation that some participants may know each other and hence could exchange information that would lead to biased data.

The platform Discord (www.discord.com) was used for the test subjects to interact with the chatbot. Discord is a platform that, after free registration, allows users to exchange content in so-called channels. There is also the option of creating own servers, which in turn are composed of the channels mentioned. The organization in servers makes it possible to create a separate area with restricted access. This possibility was one of the main criteria for choosing Discord. The platform was also favored because it offers a comparatively extensive and flexible range of functions. A single channel could be created for each test person. This made it possible to save all chat histories separately but make sure they would not be visible for the following test person. Another aspect that set Discord apart from other alternatives considered was the ability to create the fictitious chatbot as a separate user, so the message "Chatbot is typing" was displayed when interacting with the test subjects when composing a message. In total, three weeks were necessary to acquire and test all participants.

5.2.5 Measures

To measure participants' facial reactions and, thus, also their valence, Noldus FaceReader v8.1.15 (www.noldus.com) was used. FaceReader belongs to the group of so-called automated facial coding (AFC) software (Lewinski, den Uyl, & Butler, 2014). Those AFC solutions recently are developing and are attracting the interest of researchers as an alternative for, for example, facial electromyography that measures facial muscle activity and tries to link this to certain emotional reactions (Stöckli, Schulte-Mecklenbeck, Borer, & Samson, 2018). As one major drawback of facial electromyography is its high complexity when being used (Wolf, 2015), AFC software is a promising alternative and has recently been used in emotional contagion studies (e.g., Smith & Rose, 2020).

The function of FaceReader is based on research by Ekman et al. (1980) and their FACS. FaceReader measures and reports the extent of seven facial expressions. These are: happy, sad, angry, surprised, scared, disgusted, and neutral. Besides those expressions, the software analyzes the valence of the facial state. FaceReader calculates the valence by taking the happiness score and subtracting the score of the at that moment strongest negatively valenced emotion. FaceReader considers sad, angry, scared, and disgusted as negative emotions. As being surprised can both be positive and negative, this one is excluded. All scores are reported as decimal values ranging from 0 (not detected) to 1 (maximum detection).

Several authors have validated the FaceReader software in the past years (e.g., Bijlstra & Dotsch, 2011; Lewinski et al., 2014). Especially Lewinski et al. (2014) found an earlier version of FaceReader (v6) to work well if used to detect positive emotions in the form of happiness. They used the Warsaw Set of Emotional Facial Expression Pictures (Olszanowski, Pochwatko, Kukliński, Ścibor-Rylski, & Ohme, 2008) and the Amsterdam Dynamic Facial Expression Set (van der Schalk, Hawk, Fischer, & Doosje, 2011) to calculate a matching score indicating how many percent of the shown emotions were recognized correctly. Based on these findings, FaceReader appeared as a reliable tool to assess participants' facial reactions when interacting with the chatbot. To be more precise, the facial reactions were analyzed on the basis of the valence score recognized by FaceReader. Of particular interest was the delta of the valence score (i.e., the change). The change was related to the phase before the interaction with the chatbot and the actual interaction phase. The basic assumption here was that in the experimental situation, in which the chatbot displayed positive emotions, a more positive change in the valence score would have to take place than in the situation in which the chatbot did not display positive emotions. This assumption was based on the expected mimicry of positive emotions.

5.2.6 Participants

In total, 44 participants could be recruited to participate in the laboratory experiment. Of these subjects, five had to be removed because FaceReader could not recognize facial features in these cases. This was because these subjects had either a beard or wore glasses. This left a sample of 39 subjects. In total, 61.5 percent (n = 24) were male and 38.5 percent were female (n = 15). The average age was 25.18 years (SD = 4.883 years).

5.2.7 Results

Before evaluating the proposed hypothesis, the recorded videos had to be edited. The recordings started before the participant was seated to make the recording as little noticeable as possible. Furthermore, recordings were kept running even when subjects were already filling out the questionnaire following their interaction with the chatbot. This made it necessary to harmonize all videos for their analysis. Videos were recorded in.mkv format. To prevent any losses in terms of the resolution due to converting to other formats such as.mp4, videos were trimmed using LosslessCut v3.23.8.

To calculate a baseline score (i.e., a reference score of the valence without the effects of the manipulation) of the displayed emotions, the period when participants were waiting for the chatbot to start the chat was extracted. The baseline score was the mean value during this period. For comparing the two experimental groups, variables were calculated, indicating the change in the valence FaceReader could identify. Investigating the change scores was of interest, as this complied with the expectation that a facial mimicry of the displayed emotions by the chatbot must lead to a more positive change in the manipulated group compared to the control group. Change scores for valence were calculated using auxiliary regression analysis. Baseline scores (i.e., mean scores) were used as the independent variable, whereas the corresponding mean score from the phase during the video was used as the dependent variable (Chen, Ployhart, Thomas, Anderson, & Bliese, 2011). Standardized residuals were saved and used for the following t-tests. A positive (negative) value indicated that the valence was more positive (negative) relative to the baseline (i.e., facial expressions were more positive during the interaction compared to the baseline). This method of using auxiliary regression to estimate change scores is superior to simple difference score approaches because it prevents inflated errors (Schaufeli, Bakker, & Rhenen, 2009).

In line with the proposed hypothesis that expected participants to mimic the positive displayed emotions, subjects of the manipulation condition showed a more positive change of their expressed positive valence ($M_{manipulation} = .320$, $SD = .328$) compared to subjects who were assigned to the control group and were not exposed to positive displayed emotions by the chatbot ($M_{control} = -.304$, $SD = 1.283$). The t-test showed that the difference between the two investigated experimental groups was significant (95% confidence interval (CI) = -1.239 to $-.008$, $t(21.593) = -2.102$, $p < .05$). These results showed that exposing people to positive emotions displayed by a chatbot led to more positive emotions that were facially expressed by those people who interacted with the affective chatbot, compared to customers who interacted with a chatbot that did not display positive emotions. Thus, those results delivered support for the proposed H1.

5.2.8 Discussion

Results provided preliminary support for a chatbot's ability to trigger the process of primitive emotional contagion. With this, the results delivered strong support for the CASA paradigm (Reeves & Nass, 1996) and the unconscious treatment of chatbots as social entities (Katagiri et al., 2001). These findings are in line with previous research that confirmed the paradigm's applicability to interactions with AI appearing as robots (e.g., Kim et al., 2013). Even though there was no reference group featuring a human-to-human interaction, the results showed strong support for the expectation that people treat a chatbot as a social actor and the resulting behavioral patterns do equal what would be expected from human-to-human interactions. With this, the results delivered empirical support for the computing perspective (e.g., Minsky, 2006; Picard, 1995). This means, as Huang and Rust (2018) have already noted, it seems that it is less relevant whether a chatbot actually feels emotions. Instead, it is more important that it can authentically give the impression that this would be the case. This provides the first important evidence that it is wasted potential if the emotional performance of chatbots is not taken into account because too much focus is placed on their functional capabilities.

Furthermore, another central contribution made by the presented study does not only refer to emotional contagion itself but especially to the process. By relocating the experiments to a laboratory environment, the possibility was offered to record subjects during their interaction and analyze facial reactions. Adjusted for baseline values, the results indicated that subjects showed significantly more positively valenced facial expressions when interacting with the chatbot that displayed positive emotions. This indicates, in the context of AI, not only emotional contagion occurs, but the underlying process is also based on the mimicry of displayed emotions. Thus, on the one hand, the results complement prior research approaches on emotional contagion and AI. On the other hand, the results of *Study 1* support the presented findings of Lohmann et al. (2017) and Smith and Rose (2020), who found mimicry effects based on the representation of positive emotions using emojis. Thus, the empirical research delivers clearly contrary results to the assumptions of Derks et al. (2008), who excluded the occurrence of mimicry via computer-mediated communication based on speculation.

However, a significant drawback commonly associated with laboratory experiments concerns their limitations regarding the sample size, often at the expense of the complexity of the models studied due to less statistical power (Sawyer & Ball, 1981). For the present case, this meant that the focus of the first study was initially limited to the process of facial mimicry (i.e., emotional contagion). To be able to investigate a more complex research model and thus the possible influence of emotional contagion on encounter satisfaction, the research methodology was changed so that standardized video stimuli were developed, which enabled the following experiment to be conducted online. This, in turn, opened the possibility of generating a larger sample and thus investigate a more complex research question in *Study 2*.

5.3 Study 2: Mediating Role of Customer Positive Affect (Video-based Stimuli)[2]

5.3.1 Conceptual Development

In general, satisfaction with a service transaction is considered a critical outcome measure in service research (Bitner et al., 2000; Heskett et al., 1997; Zeithaml et al., 1996). In typical human-to-human interactions, it is well documented that

[2] Parts of *Study 2* have been presented at the 7th Rostock Conference on Service Research.

not only the pure functional execution of the service but also the way it is delivered are decisive for post-encounter evaluations (Oliver, 1993; Smith & Bolton, 2002; Westbrook & Oliver, 1991). In the context of emotional contagion, research has shown that a contagion of positive emotions from the employee to the customer leads to a better evaluation of the service transaction (e.g., Pugh, 2001; Tsai & Huang, 2002). The basis for this is the affect-as-information theory (Clore et al., 2001). The theory states that in situations of evaluative judgements, people use the affective state, present at that time, as a basis for their decisions. A situation in which emotional components also play a decisive role can be the post-encounter evaluation of a service (Martin et al., 2008). Martin et al. (2008) emphasize that the assessment of a service transaction is based not only on the functional component, but also on the customer's affective state at the moment of the evaluation.

The CASA paradigm assumes that people unconsciously show the same behavioral patterns and reactions when interacting with a chatbot, for example, as they would in an interaction with a human being (Reeves & Nass, 1996). On the one hand, it is the CASA paradigm that previously led to the expectation that emotional contagion from a chatbot to a human customer is possible. For service encounters, the knowledge around the CASA paradigm encourages to expect that the chatbot's affective delivery positively affects the evaluation of the service encounter. However, it is likely to expect that this relationship is mediated by the positive affective state of the customers, as is known from human-to-human interactions. On the other hand, the interaction with chatbots represents a comparatively new form of interaction for most people. This leads to a lack of experience in evaluating such a service transaction, which in turn can lead to a significant impact of the current affect on encounter satisfaction as the theory's underlying heuristics are especially beneficial when decision situations with low levels of knowledge are present (Clore et al., 2001). Furthermore, customer satisfaction in AI-based service encounters is still a relatively unexplored field (Blut et al., 2021). As customer satisfaction significantly influenced behavioral intentions, such as the intention to return (loyalty intentions) and to engage in word-of-mouth (Athanassopoulos et al., 2001), it was considered an appropriate variable to assess the functional and social performance of the chatbot.

In summary, *Study 2* was based on the expectation that in a human-to-chatbot interaction, emotional contagion is possible, which ultimately leads to a more positive affective state among interacting customers. This assumption was encouraged by the extant research and by the findings of *Study 1*. Therefore, a positive effect of positive emotions displayed by the chatbot on customer positive affect was expected that in the further course positively affected the assessment of the service experienced (Figure 5.3).

> **H2**: The effect of a chatbot's positive displayed emotions on customers' satisfaction with the service encounter is mediated by customers' positive affect.

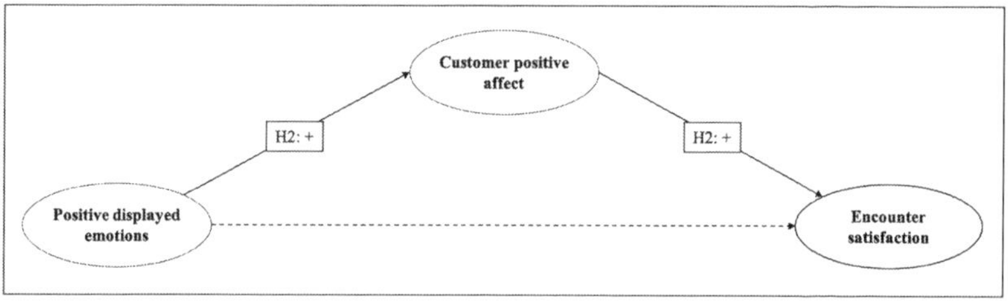

Figure 5.3 Proposed Research Model and Hypothesis, Study 2

5.3.2 Research Objectives and Setting

Study 2 had the primary goal to overcome the limitations of the laboratory experiment (*Study 1*) regarding the sample size and thus the complexity of the research model. This should help to further investigate the contagion process on the one hand and on the other hand to shed light on its influence on the outcome measure encounter satisfaction. Furthermore, the laboratory experiment had the advantage of a high external validity as customers were reliably led to believe they were interacting with a real chatbot (Homburg, 2018). However, this was at the expense of internal validity that should be more of a concern in *Study 2*. These intended objectives made it necessary to standardize the stimuli of the study to enable a larger number of subjects to participate. In the end, video-based stimuli were chosen, which should show a fictitious service chat with a chatbot (see Section 5.3.3 for a detailed explanation). The basis for the development of these videos was the hospitality industry setting from *Study 1* and, in addition, the chats that took place as part of the Wizard of Oz method. The way *Study 2* was designed had the additional advantage that a mutual validation of the results was possible. This especially concerned measuring customers' emotional reactions. While this relied on the analysis of the unconscious facial reactions of the subjects in *Study 1*, it was measured using self-report scales, which were filled out by the subjects after the experiment, in *Study 2.*

5.3.3 Experimental Design

Compared to the first study, *Study 2* was designed as a scenario-based web experiment. Standardized videos served as manipulations, each representing an experimental condition (Figure 5.4). For the experiment, participants were requested to put themselves in the customer's perspective. One main upside of standardizing all stimuli was its improvement in control over the manipulation and internal validity (Pascual-Leone, Herpertz, & Kramer, 2016). This makes video stimuli a popular tool in experimental research (e.g., Rosenthal-von der Pütten et al., 2013). Furthermore, the video stimuli allowed to include a larger sample and thus to extend the investigated research model. The experimental design again was a 2×1 between-subjects design. Participants were assigned randomly to one of the two experimental conditions. The manipulated factor was displayed emotions. For the design of those videos, the findings from the previously reported pre-studies as well, as the procedure from *Study 1*, served as point of orientation.

Following *Study 1*, a fictive booking encounter of a hotel room was used as a storyboard. The chat records from the first study indicated how a typical conversation looks like and thus served as basis. Typical components extracted included, for example, the exchange of information regarding leisure activities. Displaying a whole service encounter was to give the stimuli a higher degree of realism and to be able to design videos that lasted long enough for emotional contagion to occur (Barger & Grandey, 2006). In this regard, Shum et al. (2018) suggest a minimum of ten message turns between a human and a chatbot to lay the foundation for chatbots to prove themselves as social entities. Every interaction started with the chatbot coming online and entering the chat. This was indicated by a status indicator in the upper right corner of the display. Shortly after the chatbot was online, it started typing the first message welcoming the customer to the chat. Answering times of the chatbot and the interacting customer were sped up. However, chatbot responses were not instantly sent after a customer's message arrived. Instead, the fictive chatbot had a natural response time as the extant research shows positive impacts on the chatbot's perception (Holtgraves & Han, 2007; Moon, 1999; Schanke et al., 2021). Furthermore, the decision was encouraged by the results of the presented pre-studies, which showed how shortening of the response time negatively affected the perception of the displayed emotions.

Figure 5.4 Extracted Part of the Manipulation of Emotions for Studies 2, 4, and 5. (**Notes:** Chat protocols have been translated from German for this figure.)

To prevent subjects from abandoning the experiment because nothing happening on the screen, typing the message was indicated with "Chatbot is typing," as suggested by Gnewuch et al. (2018). Other aspects concerning the design and the behavior (e.g., gender) of the chatbot followed *Study 1*. The fictive interaction in total lasted two minutes and ten seconds.

To make sure all subjects would be carefully instructed about how to behave during the experiment, opening credits were designed. They contained information concerning the following interaction topic, the task to take the interacting customer's perspective, the requirement to watch the video until the end, and how to react in case of quality issues. Opening credits were shown before all stimuli and were not skippable. The developed videos had a size of 1920 pixels (height) by 1080 pixels (width), which enabled an easy full-screen playback on desktop computers and mobile devices. They were created using Adobe After

Effects v16.1.1 for MacOS. Before the actual experiment, the videos were shown to several people of different sexes, ages, and education levels to validate they perceive the videos and the storyboard as realistic.

5.3.4 Procedure

The experiment was conducted web-based. Participants were recruited on the crowd-sourcing website Clickworker. They received monetary compensation for their participation in the experiment. Conducting a data collection with quasi-professional participants included various risks (e.g., lack of attention or motivation), which were addressed by different a priori remedies. For example, participants received financial compensation. In addition, it was explicitly stated at the beginning of the experiment that attention checks would occur during the experiment and that the financial compensation would only be paid if these were answered correctly (Hulland & Miller, 2018). The survey was created on the platform SoSci, which offers a high degree of freedom to individualize surveys. Clickworkers were invited by providing them the link to the externally connected survey. When they clicked on it, they were first welcomed, giving them information on the experiment's purpose. They were told it dealt with the use of chatbots in service interactions. Furthermore, they were asked to take the perspective of the customer interacting with the chatbot. By clicking the "next" button, they were randomly assigned to one of the two experimental conditions described above. As Clickworkers are paid by entering a code if they had to participate in a survey that uses an external tool for the survey, as it was the case in this survey, it was ensured that they were not able to skip the video by fading-in the next button right at the moment when the video ended. The pages that followed contained the scales described below. Clickworkers were to answer either imagining to be the interacting customer (customer positive affect, encounter satisfaction, and positivity of displayed emotions) or asking for the subject's personal characteristics (experience with live chats and chatbots, need for interaction, emotional decision behavior, and demographic information). Customers' positive affect was measured at one point immediately after the fictive encounter using self-reports. Real-world experiments on emotional contagion sometimes measure participants' affective state at two points in time (i.e., before and after presenting the stimuli) (e.g., Hennig-Thurau et al., 2006). For the present study, however, this procedure appeared too fraught with the risk of social desirability due to the short time the web-based experiment lasted. Instead, the pre-encounter affective state was expected to be equal across the experimental groups due to the random assignment to the different scenarios. Thus, differences in terms of post-encounter affective state must be due to the manipulations. This procedure is standard in experimental research (e.g., Barger & Grandey, 2006).

5.3.5 Measures

Only scales that previously had already been part of related research approaches have been used to measure the different variables. Measuring the affective state using self-report scales is in line with previous research on emotional contagion (e.g., Barger & Grandey, 2006; Pugh, 2001). In experiments, especially when other techniques are not applicable, self-reports belong to the most commonly used ways to measure emotions (Kim & Moon, 1998). All scales have been translated into German, trying to stay as close as possible to the original scale but simultaneously ensuring decent language and easy comprehension. All items were measured using seven-point Likert-type scales. These ranged from "completely disagree" (1) to "agree completely" (7). The scales are described in the same order as they appeared during the experiment. Scale scores were either produced computing a weighted mean using factor loadings from the confirmatory factor analysis or were extracted from SmartPLS.

Customer positive affect

To measure customer positive affect, the positive subscale of the Job-Affect-Scale (JAS) was used. The JAS was initially developed by Burke, Brief, George, Roberson, and Webster (1989) and later used in different service-oriented studies (e.g., Hennig-Thurau et al., 2006; Pugh, 2001). According to Pugh (2001), the JAS is considered to be more appropriate for measuring affect in a business context compared to the also well-known Positive Affect Negative Affect Scale (PANAS) (Watson, Clark, & Tellegen, 1988). The JAS has been used in several studies (e.g., Allen, Pugh, Grandey, & Groth, 2010). The items were preceded by the prefix "I feel [...]". The items used were "strong," "elated," "excited," "enthusiastic," "active," and "peppy" (CA = .922, CR = .939, AVE = .720) (see Table 5.2 for a list of all items used including their factor loadings).

Encounter satisfaction

The extant literature shows that emotional contagion impacts customers' evaluation of the service experienced (e.g., Barger & Grandey, 2006; Pugh, 2001). A scale developed by Hennig-Thurau et al. (2006) was used to measure satisfaction with the transaction, as it proved to work well in emotional contagion research.

The scale was chosen over other well-known scales, for example, SERVQUAL (Parasuraman, Zeithaml, & Berry, 1988), that are used to assess service quality because it does primarily focus on a single service encounter rather than overall service performance. This was considered important as subjects only experienced a short service encounter without additional information on the company or other aspects. Using the SERVQUAL scale would have meant either having to extend the information given in the stimuli, which would have added disturbances, or having to adapt the scale such that it suits the transaction focus of this investigation. Items of the encounter satisfaction scale were, for example, "It was a good service experience" or "I am satisfied with the experienced service" (CA = .942, CR = .956, AVE = .813).

Positivity of displayed emotions

Perceived positivity of the chatbot's displayed emotions was measured to check for the intended manipulation. The positivity of the emotions was assessed with a scale derived from the JAS (see measurement of customer positive affect). Participants were asked to evaluate the emotions the chatbot showed. Compared to the JAS, the prefix was changed to "The emotions displayed by the chatbot were […]" as well as the first item that was changed from "strong" to "positive" (CA = .956, CR = .964, AVE = .819).

Experience with live chats and chatbots

The use of emojis is a common way to communicate emotions via chat applications. Therefore, participants' experience with live chats and with chatbots was measured. During the investigation, participants received the instruction that live chats do also include applications such as WhatsApp. The decisive factor was that an essential part of communication via such applications involves the use of emojis (Schlobinski & Siever, 2018). An item initially used by Gefen and Straub (2004) was the basis for developing one item to measure experience with live chat and one to measure the experience with chatbots as communication partner. The items were "I am familiar with communicating via live chats" and "I am familiar with communicating with chatbots." Since experience with communicating via live chats is not necessarily related to being experienced with communicating with chatbots, the two items were each used to form single-item variables.

Need for interaction

The extant research results clearly show significant reservations about the use of chatbots in service encounters (e.g., Mozafari et al., 2020). These reservations or the preference for human contact can lead to customers opting for human-based service in situations where customers are free to choose between using a service offered by a human compared to the same service delivered by a chatbot (e.g., Dietvorst et al., 2015) or technologies in general (e.g., Walker, Craig-Lees, Hecker, & Francis, 2002). In the present case, however, subjects did not have a choice. Since it could not be ruled out that the reservations may have an influence, for example, on encounter satisfaction, subjects' need for interaction was assessed. Need for interaction is defined as "[…] the extent to which personal contact is perceived to be needed or preferred." (Walker & Johnson, 2006, p. 127) To measure need for interaction, a scale by Walker and Johnson (2006) was used. They used this to capture the propensity of customers for face-to-face and non-computer-mediated contact. The items were adapted to capture the preference for interaction with a human and service delivery by a human. The items used were, for example, "I rather like to communicate with people when services are provided" or "I prefer personal contact for asking and answering my questions" (CA = .895, CR = .935, AVE = .827).

Emotional decision behavior

The affect-as-information theory presented above claims that the currently present affect may affect how decisions are made. Therefore, it is like to expect effects depending on how participants report how much they think they are influenced by emotional components when making decisions. For this reason, a scale by Shiv and Fedorikhin (1999) was used to evaluate whether subjects consider themselves to be guided by affect or ratio in terms of their decision making. All items were introduced with the sentence "In general, in decision making, I tend to be guided by […]." The items used were: "my rational side (1)/my emotional side (7)" and "my head (1)/my heart (7)" (CA = .883, CR = .945, AVE = .895).

Table 5.2 Items and Indicator Loadings for Latent Constructs, Study 2

Construct	Item loading	CA	CR	AVE
Main constructs				
Customer positive affect		.922	.939	.720
I feel strong.	.798			
I feel elated.	.840			
I feel excited.	.866			
I feel enthusiastic.	.866			

(continued)

Table 5.2 (continued)

Construct	Item loading	CA	CR	AVE
I feel active.	.880			
I feel peppy.	.840			
Encounter satisfaction		.942	.956	.813
I am delighted by the experienced service.	.900			
The service helped me to solve my concern.	.851			
It was a good service experience.	.940			
I am satisfied with the experienced service.	.936			
I really liked the service experience.	.878			
Controls				
Need for interaction		.895	.935	.827
I feel more relaxed when I have personal contact with service employees.	.914			
I rather like to communicate with people when services are provided.	.916			
I prefer personal contact for asking and answering my questions.	.898			
Emotional decision behavior		.883	.945	.895
In general, in decision making, I tend to be guided by my rational side/my emotional side.	.950			
In general, in decision making, I tend to be guided by my head/my heart.	.942			

Notes: CA = Cronbach's alpha, CR = composite reliability, AVE = average variance extracted

5.3.6 Participants

The original sample consisted of a total of 309 participants. Subjects were deleted from the original sample if they did not complete the survey (n = 4), if they failed to answer the attention check correctly (n = 30), or if they appeared to not have recognized the experimental treatment (n = 55) (see treatment check in Section 5.3.7.2). This procedure resulted in a final sample of 220 valid responses. Of those, 53.2 percent were male (n = 117) and 46.8 percent female (n = 103). On average, participants were 35.72 years old (SD = 11.564 years).

5.3.7 Results

5.3.7.1 Common-method Bias

The problem of common method variance (CMV) arises because of the potential bias introduced by using the same measurement methods for both independent and dependent variables. This can have inflating or deflating effects on the observed relationships (Podsakoff, MacKenzie, Lee, & Podsakoff, 2003). In the present thesis, the problem of CMV was addressed in several ways. These included so-called a priori remedies as well as statistical methods that specifically addressed CMV post hoc (Hulland, Baumgartner, & Smith, 2018; Podsakoff, MacKenzie, & Podsakoff, 2012). First, several procedural remedies were employed when designing the questionnaire to reduce CMV resulting from social desirability, consistency motifs, and implicit theories (Podsakoff et al., 2003). Participants were informed that their responses were important for research, that their opinion alone counted in answering the questions, and that no experience or specialized knowledge was a prerequisite for their participation. In addition, participants received monetary compensation to increase their motivation. Furthermore, the scales for measuring the independent and dependent variables were placed on different pages of the following questionnaire to prevent participants from editing answers to correspond to an implicit theory. Second, the problem of CMV was also addressed by applying post hoc procedures. For this purpose, a marker variable was inserted into the model that was theoretically unrelated to the other constructs but measured on the same scale (Hulland et al., 2018; Lindell & Whitney, 2001). Two items adapted from Bowling and Hammond (2008) were used as a marker variable (Hulland et al., 2018). The items used were "Overall, I am satisfied with my job" and "I like working where I am currently employed." As with the other scales, responses were measured using a seven-point Likert-type scale. Chin, Thatcher, Wright, and Steel (2013) suggest the construct level correction (CLC) approach for applying the marker variable procedure in PLS analysis. This approach is based on adding one separate marker variable to the model for each latent construct. Ultimately, this procedure removes potential CMV effects.

5.3.7.2 Treatment Check

To check whether the emotions shown by the chatbot were perceived by the subjects as positive in the manipulated condition and less positive in the control condition, the thesis follows the proposed terminology of Sigall and Mills (1998). In this context, they speak of a treatment check. The original sample, corrected in terms of subjects whose answers were incomplete or who did not correctly answer the mentioned attention check, contained 275 subjects. From this sample, subjects were removed if their responses indicated that they had not recognized the experimental treatment. On the one hand, the main underlying question was whether subjects had perceived and rated the displayed emotions as positive when watching the videos with positive emotions. On the other hand, subjects should not have perceived the emotions as positive in the control condition. In the course of the presented research design, however, it may happen that the emotions presented in the context of the described video stimuli are not perceived by the subjects as intended. This can be caused by various factors, such as inattention. The treatment check was designed to identify and exclude these subjects from the sample. This is a common procedure in experimental research (Geuens & De Pelsmacker, 2017; Perdue & Summers, 1986). Thus, as a first step, subjects' distribution concerning their attributed positivity of the displayed emotions was examined. This step revealed that some participants indicated to have perceived the emotions during the control scenario as extraordinarily positive. In contrast, other participants perceived the positive condition as being low in positivity. Consequently, subjects were removed from the sample if their positivity score was extraordinarily high if they were part of the control group (no emotions). The same procedure was applied to participants of the manipulated group assigned to the video showing positive emotions, this time, however, checking if their positivity score was extraordinarily low. With this procedure, 55 participants were deleted from the sample, resulting in a total of 220 responses that were used for further investigations and the following hypothesis testing. Before applying the described procedure, a t-test was calculated to see if, without this procedure, the emotions from the manipulated condition were perceived as significantly more positive, indicating that the intended treatment had been generally recognized, ruling out possible mistakes in the conceptualization of the stimuli. This t-test showed that the attributed positivity was already significantly different across the two experimental groups ($M_{control} = 4.601$, $SD = 1.447$, $M_{manipulation} = 5.246$, $SD = 1.180$, 95% CI = −.958 to −.332, $t(264.491) = -4.055$, $p < .001$). Thus, the procedure was considered adequate and not biasing the results.

5.3.7.3 Path Model Estimation

To test the proposed hypothesis, SmartPLS v3.3.3 was used (Ringle, Wende, & Becker, 2015). SmartPLS builds on structural equation modeling (SEM) and applies the partial least squares (PLS) procedure (Lohmoeller, 1989). In recent years applying SEM in research became more and more popular (Nitzl, Roldán, & Cepeda Carrión, 2016). One key upside of the PLS procedure is its statistical power and ability to provide a reliable and valid analysis for smaller sample sizes

(Reinartz, Haenlein, & Henseler, 2009). Furthermore, the chosen PLS method is exceptionally robust against normal distribution violations, a common research problem (Hair, Hult, Ringle, & Sarstedt, 2014a). To conduct inference tests, SmartPLS uses a bootstrapping procedure. Bootstrapping builds on repetitively building new samples with subjects of the existing sample (Hayes, 2017). In doing so, the cases of the original sample are multiple times taken to form a new sample with the same number of cases. This procedure makes bootstrapping especially more independent against violations against normal distribution by calculating an empirical representation of the sampling distribution (Wood, 2005). Inference tests were calculated using 5,000 resamplings of the used sample. In the path analysis, positive displayed emotions were operationalized as a dichotomous variable (no emotions = 0, positive displayed emotions = 1). All other variables were specified as latent constructs (consisting of the items presented in Table 5.2).

5.3.7.4 Measurement Model

Hair, Sarstedt, Ringle, and Mena (2012) suggest a two-step approach when applying PLS-SEM. Before analyzing the inner model, they recommend assessing the outer model, that is, the measurement model. This measurement model is especially important when dealing with reflective indicators, as is the case with the present investigation.

The first step in analyzing the outer model is assessing the used indicators' reliability (Hair et al., 2012). This was evaluated using the calculated factor loadings from SmartPLS. As presented in Table 5.2, all factor loadings were above or equal to .798 and hence exceeded the suggested threshold of .70 (Hulland, 1999).

The second step in evaluating the outer model is examining the constructs' internal consistency reliability (Hair et al., 2012). The most common indicator used to check for construct reliability is Cronbach's alpha (Cronbach & Meehl, 1955). However, Cronbach's alpha is based on unweighted indicators, making it a rather imprecise measure to assess the reliability of the used indicators (Hair et al., 2012). This disadvantage comes into play, especially when different factor loadings are present for a composite (Cho, 2016). For this reason, construct reliability was assessed by consulting the composite reliability (CR) (Bagozzi & Yi, 1988). All constructs demonstrated good reliability scores, as they all surpassed the suggested threshold of .70 for CR (Bagozzi & Yi, 1988) (Table 5.2).

As a third step in evaluating the outer model, Hair et al. (2012) suggest assessing convergent validity. The average variance extracted (AVE) was calculated to assess convergent validity. AVE can be considered an indicator of the variance of a latent variable that the used indicators can explain. Values above .50 mean that more variance is explained by the indicators than remains unexplained. Thus, the AVE should be at least above .50 (Bagozzi & Yi, 1988). Results indicate the AVE of each variable was at least equal to .720 (Table 5.2). Therefore, convergent validity could be approved.

As a fourth step, discriminant validity was assessed. Discriminant validity is evaluated to ensure that the measured constructs are empirically distinct from each other (Hair, Sarstedt, Hopkins, & Kuppelwieser, 2014b). To assess discriminant validity, first, the cross loadings of the indicators were examined. Based on the cross loadings, discriminant validity can be assumed if an indicator's loading on the intended latent variable is higher than its loadings on every other latent variable (cross loading) in the model (Chin, 1998; Grégoire & Fisher, 2006). As presented in Table 5.3, every indicator loaded highest on the intended latent variable. As a second step and as suggested by Hair et al. (2012), the Fornell-Larcker criterion (Fornell & Larcker, 1981) was consulted. According to the Fornell-Larcker criterion, discriminant validity is approved if the AVE's square root exceeds the correlations with every other construct in the model. As presented in the table below (Table 5.4), the square roots of the AVEs exceeded the correlation between customer positive affect and encounter satisfaction. Thus, discriminant validity, based on the procedure suggested by Fornell and Larcker (1981), could be assumed. However, the Fornell-Larcker criterion is increasingly questioned in literature (Henseler, Ringle, & Sarstedt, 2015). This is mainly because variance-based SEM, such as SmartPLS, tends to overestimate the factor loadings that are, in the further course, used to calculate the AVE (Hui & Wold, 1982; Lohmoeller, 1989). Furthermore, variance-based SEM also tends to underestimate the correlations between the constructs in a model (Marcoulides, Chin, & Saunders, 2012; Reinartz et al., 2009). Therefore, discriminant validity was alternatively assessed using the heterotrait-monotrait (HTMT) method, suggested by Henseler et al. (2015). The HTMT method compares the average heterotrait-heteromethod correlations, the correlations of the indicators of a latent variable with the indicators of another latent variable, to the monotrait-heteromethod correlations, the correlation of the items of one specific variable. For this thesis, the restrictive and conservative approach to use the method as a criterion for discriminant validity is followed (Henseler et al., 2015). Following the criterion approach, there is no common threshold that has been agreed upon. However, .85 is currently the most conservative one (e.g., Kline, 2011). When used as a criterion and compared to a threshold, discriminant validity can be assumed when the HTMT ratio is below .85. Based on the conducted analysis that resulted in a maximum ratio of .644, the conclusion could be drawn that discriminant validity was existent (Table 5.5).

Table 5.3 Discriminant Validity (Cross Loadings), Study 2

	PDE	CPA	ES	NFI	EDB	EXPLC	EXPCB	Age	Gender
PDE	*1.000*	.312	.173	−.095	.058	.058	.082	.006	.036
CPA_01	.166	*.798*	.547	−.079	.051	.226	.210	.134	−.007
CPA_02	.256	*.840*	.502	−.049	.177	.242	.245	−.004	.077
CPA_03	.245	*.866*	.583	−.122	.183	.179	.182	−.132	.045
CPA_04	.296	*.866*	.522	−.119	.168	.129	.148	−.188	.096
CPA_05	.315	*.880*	.489	−.101	.139	.151	.227	−.029	.092
CPA_06	.308	*.840*	.436	−.015	.177	.132	.220	−.027	.030
ES_01	.123	.588	*.900*	−.213	.039	.160	.135	−.005	.075
ES_02	.157	.471	*.851*	−.180	−.060	.253	.125	.032	.081
ES_03	.173	.525	*.940*	−.207	−.006	.196	.105	.019	.072
ES_04	.131	.502	*.936*	−.180	−.006	.230	.083	.009	.073
ES_05	.193	.624	*.878*	−.251	.091	.179	.165	−.042	.061
NFI_01	−.115	−.103	−.215	*.914*	.039	−.036	−.098	.009	−.026
NFI_02	−.076	−.077	−.210	*.916*	.094	−.017	−.077	.032	−.099
NFI_03	−.063	−.081	−.203	*.898*	.043	−.014	−.051	.096	−.103
EDB_01	.040	.182	.031	.061	*.950*	.023	−.058	−.001	.158
EDB_02	.071	.151	−.002	.060	*.942*	.020	−.057	−.048	.119
EXPLC	.058	.208	.224	−.025	.023	*1.000*	.591	−.089	.037
EXPCB	.082	.241	.138	−.084	−.061	.591	*1.000*	−.117	−.041
Age	.006	−.051	.001	.048	−.025	−.089	−.117	*1.000*	.070
Gender	.036	.066	.080	−.081	.147	.037	−.041	.070	*1.000*

Notes: PDE = positive displayed emotions, CPA = customer positive affect, ES = encounter satisfaction, NFI = need for interaction, EDB = emotional decision behavior, EXPLC = experience with live chats, EXPCB = experience with chatbots

Table 5.4 Discriminant Validity (Fornell-Larcker Criterion), Study 2

Latent construct		1	2	3	4
1	Customer positive affect				
2	Encounter satisfaction	.606			
3	Need for interaction	−.097	−.231		
4	Emotional decision behavior	.177	.016	.064	
Square root of the AVE		.849	.902	.909	.946

Notes: Scores indicate the correlation between the constructs

Table 5.5 Discriminant Validity (HTMT), Study 2

		1	2	3	4	5	6	7	8	9
1	Positive displayed emotions									
2	Customer positive affect	.325								
3	Encounter satisfaction	.177	.644							
4	Need for interaction	.098	.104	.249						
5	Emotional decision behavior	.063	.194	.054	.073					
6	Experience with live chats	.058	.217	.233	.026	.024				
7	Experience with chatbots	.082	.252	.140	.088	.064	.591			
8	Age	.006	.105	.024	.053	.028	.089	.117		
9	Gender	.036	.071	.083	.088	.156	.037	.041	.070	

Table 5.6 Descriptives and Correlations, Study 2

		M/%	SD	1	2	3	4	5	6	7	8	9
1	Positive displayed emotions	49.5%	-									
2	Customer positive affect	5.287	1.032	.312								
3	Encounter satisfaction	6.107	1.003	.173	.606							
4	Need for interaction	5.335	1.380	−.095	−.097	−.231						
5	Emotional decision behavior	3.746	1.462	.058	.177	.016	.064					
6	Experience with live chats	5.736	1.521	.058	.208	.224	−.025	.023				
7	Experience with chatbots	4.886	1.826	.082	.241	.138	−.084	−.061	.591			
8	Age	35.720	11.564	.006	−.051	.001	.048	−.025	−.089	−.117		
9	Gender (female)	46.8%	-	.036	.066	.080	−.081	.147	.037	−.041	.070	

Notes: Correlations equal to or above |.138| are statistically significant ($p < .05$, two-tailed).

5.3.7.5 Structural Model

Three criteria were consulted to examine the structural quality of the proposed model. As a first step, the explained variance of the endogenous variables in the model was evaluated, consulting the coefficient of determination (R^2) (Chin, 2010). This metric indicates how much variance of the endogenous variables can be explained by the predicting independent variables and thus the model structure. The illustration presented below shows that the R^2 of the variable customer positive affect was .195, whereas for encounter satisfaction it was .428 (Figure 5.5). This indicates that a substantial portion of the variance of the variable encounter satisfaction could be accounted for.

As a second step, Q^2 was calculated. As suggested by Stone (1974) and Geisser (1974), Q^2 serves as an indicator for predictive relevance. It is based on the blindfolding procedure that reuses samples to predict original values by systematically removing data points (Hair et al., 2014a; Hair et al., 2012). For the presented model, all Q^2 values were above or equal to .134 and thus above 0, which serves as a threshold (Chin, 1998; Henseler, Ringle, & Sinkovics, 2009). Thus, predictive relevance could be confirmed. See Figure 5.5 for a complete overview.

Especially the fact that SEM makes it easy to include multiple constructs in the model triggers the potential of high correlations among predicting constructs (Grewal, Cote, & Baumgartner, 2004). This overlapping of constructs in an underlying model is referred to as collinearity if two constructs are involved or multicollinearity if multiple constructs are affected (Hair, Babin, Anderson, & Black, 2018a). Collinearity harbors the risk of biasing the results in a regression analysis (Hair, Risher, Sarstedt, & Ringle, 2019). Therefore, as a third and last step, the structural model was assessed concerning collinearity. To assess collinearity, the variance inflation factors (VIF) were consulted. Hair et al. (2019) suggest collinearity as being of no concern if the VIFs are below or close to 3. For the proposed model, the VIFs were all equal to or below 1.655. From this, the conclusion was drawn that collinearity was not present. Table 5.7 presents all VIFs.

Table 5.7 Variance Inflation Factors, Study 2

		2	3
1	Positive displayed emotions	1.031	1.119
2	Customer positive affect		1.242
3	Encounter satisfaction		

(continued)

Table 5.7 (continued)

		2	3
	Need for interaction	1.033	1.038
	Emotional decision behavior	1.052	1.082
	Experience with live chats	1.558	1.568
	Experience with chatbots	1.625	1.655
	Age	1.027	1.027
	Gender	1.058	1.058

5.3.7.6 Hypothesis Testing

H2 proposed customer positive affect mediating the relationship between positive displayed emotions and encounter satisfaction in a chatbot-handled service interaction. In general, the results, which are presented in Figure 5.5, supported the derived hypothesis. The display of positive emotions had a positive and significant effect on customer positive affect ($\beta = .265$, $p < .001$). In turn, customer positive affect had significantly affected encounter satisfaction ($\beta = .602$, $p < .001$). The decisive factor for the acceptance or rejection of the hypothesis was, however, whether the indirect effect through customer positive affect had the correct mathematical sign and was statistically significant. The indirect effect of positive displayed emotions on encounter satisfaction through customer positive affect was also statistically significant (positive displayed emotions ➜ customer positive affect ➜ encounter satisfaction = .160, $p < .001$). That is, customer positive affect mediates the effect of positive displayed emotions on encounter satisfaction. As the direct effect of positive displayed emotions on encounter satisfaction lost its statistical significance when the mediator was introduced ($\beta = -.032$, $p = .541$), the results indicated a full mediation, thus delivering support for H2. All paths, including the coefficients for the contained control variables (need for interaction, emotional decision behavior, experience with live chats, experience with chatbots, age, and gender), are reported in Table 5.8.

To complement the path coefficients reported above, which formed the basis for testing the hypothesis, the so-called f^2 effect sizes were calculated (Cohen, 1988). Cohen (1988) suggests thresholds of .02, .15, and .35 to classify effect sizes as weak, moderate, and strong. This serves as a parameter for the relative impact of a specific exogenous variable on the examined model's endogenous variables. High scores indicate high relevance of the corresponding exogenous variable (Hair et al., 2018a). The f^2 effect sizes overlap with the path coefficients in their explanatory power (Hair et al., 2019) but can be used as a supplement,

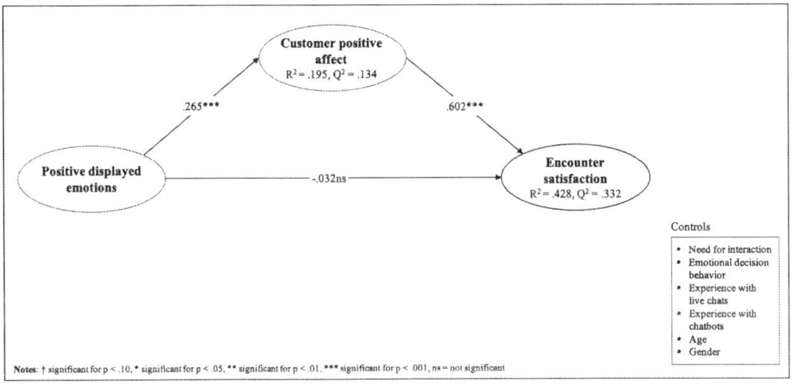

Figure 5.5 Results Structural Equation Modeling, Study 2

especially in mediation models (Hair et al., 2012; Nitzl et al., 2016). Based on the suggestions by Cohen (1988) the results indicated a weak effect size of positive displayed emotions on customer positive affect ($f^2 = .085$). In the further course, customer positive affect had a strong effect on encounter satisfaction ($f^2 = .511$). Hence, these results further supported the mediation hypothesis. Table 5.9 contains all f^2 effect sizes.

After testing the proposed hypothesis, two more t-tests were run to analyze and visualize the differences between the two experimental conditions. To do this, the unstandardized latent variable scores from SmartPLS were extracted to calculate the mean scores for the investigated variables. The first t-test showed that the control group scored significantly lower in terms of their reported positive affect ($M_{control} = 4.969$, $SD = 1.050$) compared to the manipulated group ($M_{manipulation} = 5.611$, $SD = .910$, 95% $CI = -.903$ to $-.381$, $t(214.696) = -4.848$, $p < .001$). Similar results appeared comparing the two groups concerning their encounter satisfaction. Again, the control group ($M_{control} = 5.935$, $SD = 1.124$) scored significantly lower compared to the manipulated group ($M_{manipulation} = 6.281$, $SD = .831$, 95% $CI = -.608$ to $-.083$, $t(202.613) = -2.597$, $p < .05$).

Both the results from the computed t-tests and the results from the mediation analysis showed that the display of positive emotions by a chatbot during a service encounter impacts the reported positive affect of interacting customers. As expected, a more positive reported affective state due to the display of positive emotions influences the evaluation of the service experienced, resulting in significantly better ratings in terms of encounter satisfaction. Both comparisons between the manipulated group and the control group point in the direction of the proposed hypothesis and deliver further support for a chatbot's ability to alter customers' affective state. Furthermore, they show that emotional components do also possess relevance in chatbot-handled service encounters.

Table 5.8 Standardized Path Coefficients and Significances, Study 2

	Standardized coefficient	Standard deviation	T-statistics
Main effects			
Positive displayed emotions ➜ customer positive affect	.265	.061	4.361***
Positive displayed emotions➜ encounter satisfaction	−.032	.053	.612
Customer positive affect ➜ encounter satisfaction	.602	.065	9.247***
Indirect effects			
Positive displayed emotions ➜ customer positive affect ➜ encounter satisfaction	.160	.042	3.836***
Control paths			
Need for interaction ➜ customer positive affect	−.064	.062	1.028
Need for interaction ➜ encounter satisfaction	−.176	.045	3.896***
Emotional decision behavior ➜ customer positive affect	.157	.067	2.358*
Emotional decision behavior ➜ encounter satisfaction	−.097	.056	1.736†
Experience with live chats ➜ customer positive affect	.086	.072	1.193
Experience with live chats ➜ encounter satisfaction	.175	.068	2.581*
Experience with chatbots ➜ customer positive affect	.154	.076	2.017*
Experience with chatbots ➜ encounter satisfaction	−.132	.058	2.276*
Age customer positive affect	−.014	.066	.209
Age ➜ encounter satisfaction	.040	.058	.701

(continued)

Table 5.8 (continued)

	Standardized coefficient	Standard deviation	T-statistics
Gender (female) ➔ customer positive affect	.020	.068	.292
Gender (female) ➔ encounter satisfaction	.021	.057	.363

Notes: † significant for p < .10, * significant for p < .05, ** significant for p < .01, *** significant for p < .001.

Table 5.9 Effect Sizes, Study 2

		2	3
	Main effects		
1	Positive displayed emotions	.085	.002
2	Customer positive affect		.511
3	Encounter satisfaction		
	Controls		
	Need for interaction	.005	.052
	Emotional decision behavior	.029	.015
	Experience with live chats	.006	.034
	Experience with chatbots	.018	.018
	Age	.000	.003
	Gender (female)	.000	.001

5.3.8 Discussion

The presented results from the conducted web-based experiment showed that chatbots can trigger emotional contagion in its most unconscious form, primitive emotional contagion. These results correspond to the presented results from earlier investigations on emotional contagion in an AI context (e.g., Tsai et al., 2012) and they confirm as well as extend the results of *Study 1*. Furthermore, the presented study extends extant literature by showing that even a simple type of agent that only communicates text-based possesses the ability to trigger changes in customers' affective state. In the further course, the results indicated the validity of the affect-as-information theory (Clore et al., 2001) as a positive and significant effect of customers' positive affect on encounter satisfaction was found. These findings are insofar of importance as they show that, comparable to human-based service encounters, it is not only the delivery of the service's functional components that determine its evaluation when service is provided by a chatbot. Instead, also services performed by chatbots are evaluated by customers based on their affective states and, thus, emotional components. On the one hand, this supports the already mentioned shift toward a stronger focus on the design and, thus, on chatbots' behavior (e.g., Jain et al., 2018; Liao et al., 2018). On the other hand, this shows that chatbots should, from a practical perspective, not be considered pure task performers, but their design should also consider the emotional components of service encounters.

However, *Study 1* and *Study 2* suffered from a key limitation. In both studies, subjects interacted with a chatbot or a fictitious chatbot whose responses were actually written by a human. The use of videos is a common methodology in experimental design (e.g., Rosenthal-von der Pütten et al., 2013) as is the Wizard of Oz method (e.g., Ho et al., 2018; Thies et al., 2017). However, the results of the first two studies raised the question of whether the results were indeed due to the chatbot's display of positive emotions or simply due to the comparatively high communication skills of the wizard (*Study 1*) or the videos (*Study 2*). The following *Study 3* was therefore designed to answer this very question by exchanging the chatbot's human-generated responses for a real chatbot.

5.4 Study 3: Mediating Role of Customer Positive Affect (Real Chatbot)[3]

5.4.1 Research Objectives and Setting

The first two presented studies offered, on the one hand, a high degree of external validity through the methods chosen in the form of a laboratory study with its Wizard of Oz method (Ho et al., 2018; Homburg, 2018) and, on the other hand, a high degree of internal validity through the use of standardized video stimuli (Pascual-Leone et al., 2016). The weaknesses associated with each method, for example usually small samples in laboratory experiments (Sawyer & Ball, 1981), were largely compensated for by the other method. In addition, the emotional responses of the subjects in both studies were measured in different ways, which both provided clear evidence that emotional contagion of positive emotions from a chatbot to a human is possible. Nevertheless, a key limitation adhered to both research approaches. In both cases, the chatbot's responses were written by humans. Although subjects were reliably led to believe that they are interacting

[3] Parts of *Study 3* have been presented at the 7th Rostock Conference on Service Research.

with a chatbot when the Wizard of Oz method was used, the interaction partner still was a human being. The same problem held true for the videos used in *Study 2*. Hence, the question was raised to what extent the shown affective processes continued to occur when a real chatbot was communicating, which may not be able to communicate at such a high level in terms of communication skills (Ho et al., 2018).

Study 3 was designed to specifically answer this question and thus validate the reliability of the existing results of the previous studies, especially the results of *Study 2*. This was related to the fact that the present study was conceptually based on the same fundament as *Study 2* and thus aimed to test the previously proposed H2 using a different methodology for the experiment. This required the programming of a real chatbot. The basis for this approach was provided by the Chatfuel platform (www.chatfuel.com), which offers the possibility to develop chatbots, for example, to use them on Facebook pages. This method, which has recently been employed in other research approaches (e.g., De Cicco, Silva, & Alparone, 2020; De Cicco, Silva, & Alparone, 2021; Liu & Sundar, 2018), is based on equipping the chatbot with predefined answers, which are then sent depending on the users' inputs. This offers the advantage that despite a high degree of flexibility in the interaction with the chatbot from the users' perspective and the associated external validity, a high degree of internal validity is still ensured.

5.4.2 Experimental Design

Following *Study 2*, *Study 3* was designed as a scenario-based web experiment. The experimental design was a 2×1 between-subjects design, where subjects were randomly assigned to one of the two scenarios. The responses of the chatbot in the video stimuli designed for *Study 2* could be used to fill the chatbot's response database. Using the known messages, the behavior of the chatbot concerning the expression of positive displayed emotions mostly resembled the presented behavior applied during the first two studies and *Study 2* in particular. Furthermore, essential characteristics of the chatbot (e.g., name and gender) could be adopted.

5.4.3 Procedure

Chatfuel allows embedding the chat window on any web page. This made it possible to conduct the study web-based. Subjects were again recruited via the crowd-sourcing website Clickworker. The same a priori remedies to ensure high data quality presented in *Study 2* were applied. Among others, this again included an attention check in the course of the questionnaire and an explicit indication that subjects would only be paid if the check was answered correctly. SoSci again served as the platform for the experiment.

To increase the degree of realism around the designed booking scenario and the programmed chatbot, the procedure applied during *Study 1* served as a point of orientation. As part of their invitation to the experiment, subjects were informed that they would be participating in the development phase of a chatbot which was currently being developed to handle the booking of hotel rooms on booking websites in the future. By clicking on the provided invitation link, subjects were given a specific task. This task included the booking of a hotel room (single or double room) for a given date in Munich. When the experiment was continued, the chat window opened, whereupon the chatbot greeted the subjects and explained its capabilities. The chats ran completely free from this point on. The end of each chat marked the booking of the hotel room specified in the task. Subjects were then able to continue the experiment on their own and automatically reached the questionnaire.

5.4.4 Participants

Initially, the study included 285 subjects who had fully completed the booking process. Of these, however, subjects had to be removed if they had not answered the subsequent questionnaire completely ($n = 23$) or had answered the included attention check incorrectly ($n = 21$). Several subjects also had to be removed because their response behavior indicated that they had not recognized the experimental treatment ($n = 48$) (see Section 5.4.5.2). This procedure resulted in a final data set containing 193 subjects. Of the subjects included, 59.1 percent were male ($n = 114$) and 40.9 percent were female ($n = 79$). The mean age was 38.31 years ($SD = 11.907$ years).

5.4.5 Results

5.4.5.1 Common-method Bias

Following the previously presented study, CMV was addressed by applying different a priori and post hoc remedies. The a priori remedies were equal to the study presented above (e.g., measuring independent and dependent constructs on different pages of the questionnaire). Furthermore, statistical remedies were applied to remove CMV effects by adding a theoretically unrelated marker variable measured on the same scale (Hulland et al., 2018; Lindell & Whitney, 2001). For the application of the CLC approach (Chin et al., 2013), the blue attitude scale (Simmering, Fuller, Richardson, Ocal, & Atinc, 2015) was taken and used to eliminate CMV effects statistically. This scale measures the attitude toward the color blue using three items: "I prefer blue to other colors," "I like the color blue," and "I like blue clothes." The answers were measured using seven-point Likert-type scales.

5.4.5.2 Treatment Check

Study 2 had shown that it is necessary to subject the data to an in-depth review regarding the experimental treatment (Sigall & Mills, 1998). In a first step, it was examined whether the intended manipulation of positive emotions was generally recognized. A t-test showed that subjects in the manipulated condition perceived the chatbot's emotions as significantly more positive ($M_{manipulation} = 4.901$, $SD = 1.398$) than subjects assigned to the control condition ($M_{control} = 4.404$, $SD = 1.445$, 95% $CI = -.857$ to $-.136$, $t(238.583) = -2.710$, $p < .01$). Thus, it could be assumed that the manipulation of positive emotions was successful. However, the data distribution indicated that some subjects showed response patterns that gave reason to believe that they had not recognized the experimental treatment. For this reason, in a second step, subjects were removed from the data set who had rated the emotions as exceptionally positive, although they were part of the control condition. Similarly, subjects were removed if they had rated the emotions extraordinarily low in positivity, even though they were part of the manipulated condition. Based on the previously significant t-test, this procedure was estimated to be appropriate and not biasing the results.

5.4.5.3 Path Model Estimation

The basic design of the study was the same as that of *Study 2*. This opened the possibility to retest H2 proposed in *Study 2* using SmartPLS v3.3.3. For inference tests, once more a bootstrapping procedure with 5,000 resamplings was relied upon. Positive emotions were operationalized as a dichotomous variable (no emotions = 0, positive emotions = 1). The other variables were specified as latent constructs, as summarized in Table 5.10.

5.4.5.4 Measurement Model

To test the measurement model, again, a four-step approach to first assess the reliability of the used indicators before controlling for internal consistency, convergent validity as well as for discriminant validity was applied.

As a first step, following the recommendations by Hair et al. (2012), no violation against the underlying threshold (.70) to assess indicator reliability could be detected (Hulland, 1999), as all indicators loaded at least .780 on their respective factor (Table 5.10). The CR of the constructs was at least .951 in all cases, indicating good internal consistency, as they surpassed .70 (Bagozzi & Yi, 1988) (Table 5.10). Assessing convergent validity, the AVEs exceeded a value of at least .763 in all cases. The AVEs were thus all above the recommended threshold of .50 (Bagozzi & Yi, 1988) (Table 5.10).

As a final step, the results were examined for possible violations against discriminant validity. Table 5.11 shows that the indicators each loaded highest on their intended factor. Consulting the Fornell-Larcker criterion (Fornell & Larcker, 1981), no violation was found either, as the square roots of the AVEs did not exceed the correlations with the other constructs in the model (Table 5.12). The HTMT method (Henseler et al., 2015) showed no violation against discriminant validity with a maximum ratio of .665 being below the conservative threshold of .85 (Kline, 2011) (Table 5.13).

Table 5.10 Items and Indicator Loadings for Latent Constructs, Study 3

Construct	Item loading	CA	CR	AVE
Main constructs				
Customer positive affect		.938	.951	.763
I feel strong.	.780			
I feel elated.	.900			
I feel excited.	.910			
I feel enthusiastic.	.903			
I feel active.	.863			
I feel peppy.	.878			
Encounter satisfaction		.943	.957	.816
I am delighted by the experienced service.	.917			
The service helped me to solve my concern.	.850			
It was a good service experience.	.947			
I am satisfied with the experienced service.	.927			
I really liked the service experience.	.872			
Controls				
Need for interaction		.924	.952	.868

(continued)

Table 5.10 (continued)

Construct	Item loading	CA	CR	AVE
I feel more relaxed when I have personal contact with service employees.	.907			
I rather like to communicate with people when services are provided.	.947			
I prefer personal contact for asking and answering my questions.	.940			
Emotional decision behavior		.924	.962	.928
In general, in decision making, I tend to be guided by my rational side/my emotional side.	.951			
In general, in decision making, I tend to be guided by my head/my heart.	.975			

Notes: CA = Cronbach's alpha, CR = composite reliability, AVE = average variance extracted

Table 5.11 Discriminant Validity (Cross Loadings), Study 3

	PDE	CPA	ES	NFI	EDB	EXPLC	EXPCB	Age	Gender
PDE	*1.000*	.273	.226	−.085	.099	.141	.210	−.215	.006
CPA_01	.127	*.780*	.530	−.039	−.023	.237	.201	−.100	−.002
CPA_02	.287	*.900*	.538	−.145	.083	.187	.165	−.190	.036
CPA_03	.290	*.910*	.502	−.194	.113	.173	.155	−.244	−.079
CPA_04	.276	*.903*	.505	−.165	.184	.107	.119	−.265	−.066
CPA_05	.182	*.863*	.370	−.010	.152	.136	.114	−.112	−.046
CPA_06	.236	*.878*	.388	−.064	.154	.116	.093	−.091	−.050
ES_01	.206	.527	*.917*	−.153	.078	.237	.241	−.069	.062
ES_02	.205	.371	*.850*	−.050	−.001	.254	.196	.031	.104
ES_03	.185	.478	*.947*	−.144	.048	.272	.193	.032	.086
ES_04	.156	.472	*.927*	−.155	.033	.247	.177	−.032	.092
ES_05	.259	.597	*.872*	−.213	.121	.152	.172	−.119	.065
NFI_01	.000	−.085	−.124	*.907*	−.101	−.072	−.106	−.002	−.022
NFI_02	−.142	−.109	−.183	*.947*	−.181	−.066	−.102	.091	−.058

(continued)

Table 5.11 (continued)

	PDE	CPA	ES	NFI	EDB	EXPLC	EXPCB	Age	Gender
NFI_03	−.075	−.158	−.143	*.940*	−.027	−.091	−.102	.060	.036
EDB_01	.109	.118	.032	−.132	*.951*	.027	.023	.048	.253
EDB_02	.086	.123	.087	−.093	*.975*	−.039	−.003	.075	.239
EXPLC	.141	.183	.254	−.082	−.012	*1.000*	.665	−.066	−.004
EXPCB	.210	.164	.217	−.111	.008	.665	*1.000*	−.045	−.145
Age	−.215	−.201	−.040	.059	.066	−.066	−.045	*1.000*	−.105
Gender	.006	−.039	.089	−.017	.254	−.004	−.145	−.105	*1.000*

Notes: PDE = positive displayed emotions, CPA = customer positive affect, ES = encounter satisfaction, NFI = need for interaction, EDB = emotional decision behavior, EXPLC = experience with live chats, EXPCB = experience with chatbots

Table 5.12 Discriminant Validity (Fornell-Larcker Criterion), Study 3

Latent construct		1	2	3	4
1	Customer positive affect				
2	Encounter satisfaction	.549			
3	Need for interaction	−.128	−.164		
4	Emotional decision behavior	.126	.066	−.113	
Square root of the AVE		.874	.903	.931	

Notes: Scores indicate the correlation between the constructs

Table 5.13 Discriminant Validity (HTMT), Study 3

		1	2	3	4	5	6	7	8	9
1	Positive displayed emotions									
2	Customer positive affect	.275								
3	Encounter satisfaction	.231	.567							
4	Need for interaction	.081	.129	.167						
5	Emotional decision behavior	.105	.144	.067	.123					
6	Experience with live chats	.141	.188	.265	.085	.035				
7	Experience with chatbots	.210	.167	.223	.115	.014	.665			
8	Age	.215	.198	.065	.057	.066	.066	.045		
9	Gender	.006	.055	.093	.043	.265	.004	.145	.105	

Table 5.14 Descriptives and Correlations, Study 3

		M/%	SD	1	2	3	4	5	6	7	8	9
1	Positive displayed emotions	50.3%	-									
2	Customer positive affect	4.917	1.190	.273								
3	Encounter satisfaction	5.542	1.194	.226	.549							
4	Need for interaction	5.194	1.358	−.085	−.128	−.164						
5	Emotional decision behavior	3.398	1.474	.099	.126	.066	−.113					
6	Experience with live chats	5.601	1.447	.141	.183	.254	−.082	−.012				
7	Experience with chatbots	5.073	1.583	.210	.164	.217	−.111	.008	.665			
8	Age	38.310	11.907	−.215	−.201	−.040	.059	.066	−.066	−.045		
9	Gender (female)	40.9%	-	.006	−.039	.089	−.017	.254	−.004	−.145	−.105	

Notes: Correlations equal to or above |.141| are statistically significant ($p < .05$, two-tailed).

5.4.5.5 Structural Model

To test the structural model, the first step was to examine the explained variance of the endogenous variables. The R^2 values of .151 for customer positive affect and .374 for encounter satisfaction showed that a good proportion of the variance could be explained by the expression of positive emotions (Figure 5.6). Based on the Q^2 values, predictive relevance of the model could be confirmed, as the values were equal to or above .104 and thus above the threshold of 0 (Chin, 1998; Henseler et al., 2009) (Figure 5.6). Since the VIFs did not exceed a value of 1.981, no violation against the threshold of 3 (Hair et al., 2019) was found, indicating that the model did not suffer from collinearity (Table 5.15).

Table 5.15 Variance Inflation Factors, Study 3

		2	3
1	Positive displayed emotions	1.112	1.158
2	Customer positive affect		1.177
3	Encounter satisfaction		
	Need for interaction	1.033	1.041
	Emotional decision behavior	1.113	1.134
	Experience with live chats	1.937	1.958
	Experience with chatbots	1.981	1.979
	Age	1.099	1.135
	Gender	1.146	1.154

5.4.5.6 Hypothesis Testing

The primary goal of the third study was to replicate *Study 2* and thus validate the results obtained through the use of video-based stimuli. In the context of *Study 2*, H2 hypothesized that customer positive affect mediates the relationship between positive displayed emotions of a chatbot and encounter satisfaction. This hypothesis could again be confirmed based on the data of the third study where subjects interacted with a real chatbot since the expression of positive emotions had a positive effect on customer positive affect ($\beta = .196$, $p < .01$), which in the further course positively influenced encounter satisfaction ($\beta = .514$, $p < .001$). The indirect effect was significant (positive displayed emotions ➜ customer positive affect ➜ encounter satisfaction = .101, $p < .01$), while the direct effect lost its significance when the mediator was introduced ($\beta = .077$, $p = .182$). Thus, the results again confirmed the existence of a full mediation and, thus, H2. All path coefficients and their significances are presented in Table 5.16.

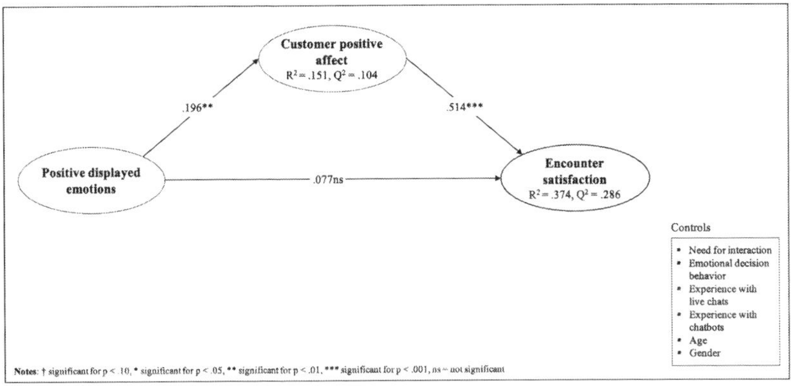

Figure 5.6 Results Structural Equation Modeling, Study 3

Based on the classification of f^2 effect sizes proposed by Cohen (1988), a weak effect of positive displayed emotions on customer positive affect was present ($f^2 = .041$). This in turn had a strong effect on customer positive affect ($f^2 = .358$). All effect sizes are presented in Table 5.17.

To visualize the differences in the means between the two experimental groups, two t-tests were calculated following the testing of the hypothesis. The first t-test compared customer's reported positive affect. Indeed, participants assigned to the manipulated condition reported more positive affect ($M_{manipulation} = 5.239$, SD = 1.024) compared to subjects assigned to the control condition ($M_{control} = 4.592$, SD = 1.261, 95% CI = −.973 to −.321, $t(182.511) = -3.911$, $p < .001$). The same pattern emerged when looking at the differences regarding encounter satisfaction. Again, participants interacting with the affective chatbot reported to be more satisfied with the service ($M_{manipulation} = 5.809$, SD = .971) than did those who interacted with the non-affective chatbot ($M_{control} = 5.272$, SD = 1.335, 95% CI = −.869 to −.205, $t(173.451) = -3.194$, $p < .01$).

Table 5.16 Standardized Path Coefficients and Significances, Study 3

	Standardized coefficient	Standard deviation	T-statistics
Main effects			
Positive displayed emotions ➔ customer positive affect	.196	.067	2.923**

(continued)

Table 5.16 (continued)

	Standardized coefficient	Standard deviation	T-statistics
Positive displayed emotions ➜ encounter satisfaction	.077	.057	1.336
Customer positive affect ➜ encounter satisfaction	.514	.073	7.029***
Indirect effects			
Positive displayed emotions ➜ customer positive affect ➜ encounter satisfaction	.101	.039	2.603**
Control paths			
Need for interaction ➜ customer positive affect	−.080	.081	.986
Need for interaction ➜ encounter satisfaction	−.091	.073	1.250
Emotional decision behavior ➜ customer positive affect	.133	.079	1.677†
Emotional decision behavior ➜ encounter satisfaction	−.058	.068	.860
Experience with live chats ➜ customer positive affect	.111	.096	1.150
Experience with live chats ➜ encounter satisfaction	.075	.078	.956
Experience with chatbots ➜ customer positive affect	.020	.097	.205
Experience with chatbots ➜ encounter satisfaction	.084	.078	1.068
Age ➜ customer positive affect	−.175	.073	2.396*
Age ➜ encounter satisfaction	.100	.062	1.617
Gender (female) ➜ customer positive affect	−.083	.080	1.032
Gender (female) ➜ encounter satisfaction	.154	.059	2.588*

Notes: † significant for $p < .10$, * significant for $p < .05$, ** significant for $p < .01$, *** significant for $p < .001$.

Table 5.17 Effect Sizes, Study 3

		2	3
	Main effects		
1	Positive displayed emotions	.041	.008
2	Customer positive affect		.358
3	Encounter satisfaction		
	Controls		
	Need for interaction	.007	.013
	Emotional decision behavior	.019	.005
	Experience with live chats	.007	.005
	Experience with chatbots	.000	.006
	Age	.033	.014
	Gender (female)	.007	.033

5.4.6 Discussion

The results of *Study 3* provided important insights for the remaining course of the thesis, especially concerning the methodology used. Specifically, this means that the interaction with a real chatbot evoked the same emotional reactions in test subjects as viewing the video stimuli in the previous study (*Study 2*) in which subjects had to mentally put themselves in the customer's perspective. *Study 3* thus fulfilled the set objective of validating the previous study's results and the chosen methodology. Furthermore, *Study 3*, which involved interaction with a real chatbot, did not come to different results than did *Study 1*, which was based on the Wizard of Oz method, or *Study 2*, in which subjects experienced a fictitious service encounter through video-based stimuli as previously already applied in other research approaches (e.g., Rosenthal-von der Pütten et al., 2013). As already indicated, this had significant influences on the further course of the thesis, since through the mutual validation the use of videos as stimuli and self-report scales was shown to be a reliable procedure. Thus, in the further course of the thesis, these methodologies (i.e., video-based stimuli and self-report scales) formed the basis for the conducted studies.

However, the focus of the studies up to this point was only on the unconscious process of emotional contagion. The process of empathy, which is characterized by more consciousness and cognitive effort, was not considered (see Section 3.2.3 for a detailed discussion). This raised the question of whether a chatbot is only able to trigger primitive emotional contagion, or if it is also able to lead customers to feel empathy toward the chatbot. This is a relevant question as people do apparently tend to unconsciously consider computers and other artificial entities as social actors, as, for example, suggested by the CASA paradigm and confirmed by the presented results. However, on a conscious level, people regularly negate the attribution of such behavioral patterns (Reeves & Nass, 1996) or adapt their behavior toward showing less effort to establish a relationship with the artificial agent (Shechtman & Horowitz, 2003). As empathy is typically considered the conscious counterpart to emotional contagion (e.g., Decety & Lamm, 2006; Prochazkova & Kret, 2017), the already existing model was expanded, adding empathy (i.e., the situational ability of being able to put yourself into the emotional state of the chatbot) as a second parallel mediation path leading from positive displayed emotions to encounter satisfaction.

5.5 Study 4: Empathy Toward Chatbots and Personality-dependent Susceptibility[4]

5.5.1 Conceptual Development

In addition to the CASA paradigm, anthropomorphism plays a central role in the interaction with nonhuman entities. It has been intensively investigated in recent years, mainly in research on robotics (e.g., Eyssel, Hegel, Horstmann, & Wagner, 2010; Hegel, Krach, Kircher, Wrede, & Sagerer, 2008; Kim et al., 2013). According to Epley et al. (2007, p. 864), anthropomorphism is "[…] the tendency to imbue the real or imagined behavior of nonhuman agents with humanlike [sic] characteristics, motivations, intentions, or emotions." This definitional approach highlights an essential distinguishing feature of anthropomorphism compared to the CASA paradigm discussed above. Because the latter is considered an unconscious process that is denied on a conscious level, the tendency of anthropomorphism reaches more strongly in the direction of a conscious attribution. This notion is illustrated by a possible explanatory approach of Guthrie (1993). This approach assumes that people use the attribution of human-like features to explain new and complex aspects. As a well-known and tangible example, Guthrie (1993) cites religion, which in many cases uses a human-like god or several human-like gods to explain such complex facts as the origin of the earth.

[4] Parts of *Study 4* have been presented at the 7th Rostock Conference on Service Research.

With the arrival of digital agents in everyday life, the demonstrated human tendency to ascribe human attributes to artificial entities such as an affective state has opened a whole new research field. Especially scholars dealing with robotics or embodied agents started intensively investigating empathy in this context. Published research in this area mainly follows two research streams: (1) emphatic reactions by the agent (e.g., Li et al., 2017; Luo et al., 2019; Schneider et al., 2012) and (2) emphatic reactions toward the agent (e.g., Kwak et al., 2013; Riek et al., 2009b; Rosenthal-von der Pütten et al., 2013; Rosenthal-von der Pütten et al., 2014). For the further course of this thesis, especially the second stream, is of importance.

The extant research on whether artificial agents can cause empathy is, however, subject to a significant limitation, as it mostly builds on triggering empathy by inflicting pain to the agent (e.g., Paiva et al., 2004; Rosenthal-von der Pütten et al., 2013; Rosenthal-von der Pütten et al., 2014), although empathy in its basic understanding is not limited to negatively valenced emotions. Existing research thus underlies a clear limitation in the context of AI. For example, Rosenthal-von der Pütten et al. (2013) were able to show that subjects reported empathic concern for a robot that was being tortured. Furthermore, they figured that this effect happened irrespective of whether they had previously interacted with the robot. Shortly after the mentioned experiment, Rosenthal-von der Pütten et al. (2014) repeated their experiment, this time including not only a human-to-robot scenario but a human-to-human scenario as well. However, they could not unveil significant differences in empathy toward either the human or the robot. As their subjects' brain activity was simultaneously scanned using fMRI, they found similar activation patterns across their used experimental conditions, regardless of participants being exposed to a human-to-human interaction or a human-to-robot interaction. With their results, the authors supported previous findings from Paiva et al. (2004), who assessed people's empathic reactions toward bullied virtual characters. In general, extant research results on empathy speak a clear language in the context of negative emotions. Thus, this encourages to assume empathy emerging also in the context of positive emotions. Of interest is not the general ability of a customer (i.e., predictive empathy) but the situation-related perspective (i.e., situational empathy). Situational empathy is generally understood as the ability to put oneself in the other person's role and perspective during an interaction (Rogers, 1959). While empathy always maintains a separation between one's own and the observed affective state, this line becomes blurred in emotional contagion. In this context, Preston and De Waal (2002) point out that the two constructs ((primitive) emotional contagion and empathy) cannot be considered independently of each other, since they share several commonalities. This statement is supported by modern research methods such as fMRI, for example, which has shown that both, to a large extent, activate the same areas of the brain (Anders et al., 2011).

From extant research results, it can be concluded that emotions seem to play a role in the interaction between a chatbot and a human, not only on a subconscious level. Instead, it can be strongly assumed that humans also consciously perceive the emotions displayed by a chatbot and thus the chatbot's related affective state. For the further course of the thesis, it is therefore expected that empathy, meaning the situational ability to empathize with the affective state of the chatbot, plays a central role as a second mediation path in the relationship between positive displayed emotions and encounter satisfaction. This means, the expression of positive emotions by the chatbot does not only lead to an unconscious change of the affective state. Instead, the expectation is that the expression of positive emotions also acts as a human-like cue (Blut et al., 2021), which leads people to attribute human features, namely an own affective state, to the chatbot. With the feeling of empathy toward the chatbot, an evaluation process runs in customers that primarily evaluates the appropriateness of the chatbot's affective delivery (Gremler & Gwinner, 2000). In the further course of this evaluation, it comes into play that it is known from human-to-human research that a positive perception of the employee by the customer has beneficial effects on customer satisfaction (Keh, Ren, Hill, & Li, 2013). Thus, in the proposed research model, empathy takes on a more cognitive and conscious role as the counterpart to emotional contagion. This means that empathy is added to the model to act as a second mediation path in the relationship between positive displayed emotions and encounter satisfaction. Therefore, the following hypothesis is proposed (Figure 5.7):

> **H3**: The effect of a chatbot's positive displayed emotions on customers' satisfaction with the service encounter is mediated by empathy.

The previous studies were able to show that a chatbot is able to cause unconscious affective customer responses through the expression of positive emotions. Based on this, it was hypothesized that the expression of positive emotions can also increase the feeling of empathy toward the chatbot (H3). However, when considering affective customer reactions, it can be assumed that not all customers react in the same way to the expressed emotions of the chatbot. A decisive factor for these different reactions can be personality traits. In this respect, human-to-human research has shown that personality traits have the potential to change reactions to emotional expressions (e.g., Shiota, Keltner, & John, 2006). In terms of personality trait research, five universal personality traits have been found to be responsible for shaping the core personality (Devaraj, Easley, & Crant, 2008). These so-called "Big Five" personality traits (extraversion, openness to experience, conscientiousness, agreeableness, and neuroticism) have already been the subject of various research approaches and their influence on affective and cognitive processes is well documented (McCrae & Terracciano, 2005).

Previous research approaches could show extraversion significantly influencing how strongly an individual displays positive emotions (Shiota et al., 2006). It could be shown that individuals who are characterized by a high degree of extraversion tend to express positive emotions more frequently and more strongly. Other researchers have found this personality trait can also lead to a stronger reaction of these individuals to positive emotions of other people (Larsen & Ketelaar, 1989). For instance, research has shown that extraverted individuals were more strongly influenced by positive emotions in movies compared to more introverted individuals (Gross, Sutton, & Ketelaar, 1998). Based on these existing research findings, it can be assumed that customers who are characterized by high levels of extraversion will unconsciously respond more strongly to the chatbot's positive emotions. This means, the affective state of these customers is more strongly influenced by the positive emotions of the chatbot.

With a stronger focus on the more conscious path of empathy, the personality trait openness to experience comes into play. This personality trait is characterized by affected individuals being more open-minded and more engaged in new activities (Griffin & Hesketh, 2004; McCrae & Costa, 1997). It should also be emphasized that individuals with a high openness to experience adapt more easily to new and changing situations. For the present context, this suggests the expectation that customers with a high openness to experience will be more open toward the affective state of the chatbot. This also implies that this openness toward the chatbot and its affective delivery should be reflected in a faster and stronger empathic reaction toward it. Accordingly, it is hypothesized that openness to experience positively moderates the effect of the chatbot's positive displayed emotions on the feeling of empathy toward it. With regard to the two personality traits discussed and their influences on the affective responses by customers, the following is therefore hypothesized (Figure 5.7):

H4: Extraversion positively moderates the effect of a chatbot's positive displayed emotions on customer positive affect.
H5: Openness to experience positively moderates the effect of a chatbot's positive displayed emotions on the feeling of empathy toward the chatbot.

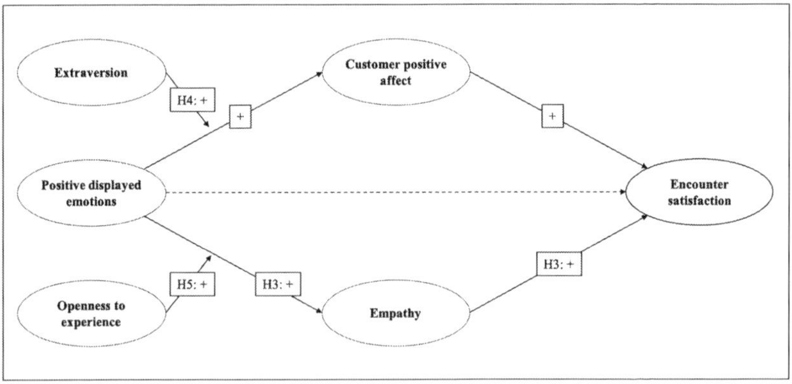

Figure 5.7 Proposed Research Model and Hypotheses, Study 4

5.5.2 Research Objectives and Setting

Study 4 served to pursue two objectives. First, the main objective was to investigate empathy as the more conscious and cognitive counterpart to emotional contagion and, with this, its role in the relationship between the affective delivery of a chatbot and encounter satisfaction. Second, the moderating role of openness to experience and extraversion on the two investigated affective customer responses should be examined. Because the previous studies gave no cause for concern regarding the chosen booking encounter (*Study 1*, *Study 2,* and *Study 3*) as well as regarding the chosen video stimuli (*Study 2*), again, a booking process of a hotel room was chosen. Since it became clear from the previous studies that the use of videos simulating a service interaction worked just as well as subjects' direct interaction either with a fictitious chatbot relying on the Wizard of Oz method (*Study 1*) or the interaction with a real chatbot (*Study 3*), videos formed the basis for the following experiment. The previous studies could also show that the measurement of the subjects' emotional reactions via self-report scales (*Study 2* and *Study 3*) did not yield different results than the measurement via facial recognition (*Study 1*). Based on these solid results, video-based stimuli and self-report scales were used for the subsequent *Study 4.*

5.5.3 Experimental Design

As before, the experimental design was a 2×1 between-subjects design. As the developed stimuli proved to work properly, the same stimuli as in *Study 2* were used to conduct the web-based experiment for *Study 3*. Similarly, the manipulated factor was positive displayed emotions.

5.5.4 Procedure

Previously, Clickworker proved to be a reliable source to gain access to many participants for a web-based survey quickly. Because of these positive experiences, participants were again recruited using Clickworker. They received monetary compensation for their participation. The setup on SoSci was identical to *Study 2*. The same a priori remedies as during *Study 2* were applied to ensure data quality. For example, the monetary compensation was used to enhance motivation. Furthermore, Clickworkers were informed at the beginning of the experiment that the survey contained attention checks, and they would only be paid if they answered those correctly. Empathy, using the scale explained below, was measured immediately after participants' positive affect. Extraversion and openness to experience were the last scales before the demographics.

5.5.5 Measures

Empathy

Not only the basic study design but also the measurement instruments used followed *Study 2* (see Table 5.18 for an overview of all constructs). However, due to the inclusion of empathy as a considered construct, it was necessary to include an additional scale for its measurement. To measure empathy toward the chatbot and its emotional state, items from two different research approaches were used to form the corresponding scale. Two items were adapted from Batson, Lishner, Cook, and Sawyer (2005). The items were "I felt compassionate toward the chatbot" and "I felt soft-hearted toward the chatbot." Another item was taken from Rosenthal-von der Pütten et al. (2013) and adapted to the chatbot context ("I was very close to the chatbot's emotions."). As before, participants were able to rate their consent or their rejection on a seven-point Likert-type scale ranging from "completely disagree" (1) to "agree completely" (7) (CA = .629, CR = .788, AVE = .553).

Extraversion and openness to experience

To measure customers' extraversion and openness to experience, the measurement inventory by Lang, John, Lüdtke, Schupp, and Wagner (2011) was used. Both personality traits were measured with two items each. For example, extraversion (CA = .694, CR = .846, AVE = .736) was measured with "I see myself as someone who is talkative" and openness to experience (CA = .566, CR = .822, AVE = .697) with "I am imaginative and come up with new ideas." For the translation into German, the work of Rammstedt and John (2005) served as a basis. The remaining three personality traits were not used as control variables. The decision was made to keep the complexity of the model at a manageable level.

Table 5.18 Items and Indicator Loadings for Latent Constructs, Study 4

Construct	Item loading	CA	CR	AVE
Main constructs				
Customer positive affect		.882	.910	.629
I feel strong.	.731			
I feel elated.	.787			
I feel excited.	.897			
I feel enthusiastic.	.767			
I feel active.	.746			
I feel peppy.	.820			
Empathy		.629	.788	.553
I was very close to the chatbot's emotions.	.723			
I felt compassionate toward the chatbot.	.747			
I felt soft-hearted toward the chatbot.	.761			
Encounter satisfaction		.888	.918	.691
I am delighted by the experienced service.	.873			
The service helped me to solve my concern.	.843			
It was a good service experience.	.855			
I am satisfied with the experienced service.	.770			
I really liked the service experience.	.814			
Extraversion		.694	.846	.736
I am outgoing, sociable.	.965			

(continued)

Table 5.18 (continued)

Construct	Item loading	CA	CR	AVE
I see myself as someone who is talkative.	.736			
Openness to experience		.566	.822	.697
I am imaginative and come up with new ideas.	.848			
I value artistic, aesthetic experiences.	.822			
Controls				
Need for interaction		.863	.915	.782
I feel more relaxed when I have personal contact with service employees.	.902			
I rather like to communicate with people when services are provided.	.866			
I prefer personal contact for asking and answering my questions.	.885			
Emotional decision behavior		.692	.852	.745
In general, in decision making, I tend to be guided by my rational side/my emotional side.	.764			
In general, in decision making, I tend to be guided by my head/my heart.	.952			

Notes: CA = Cronbach's alpha, CR = composite reliability, AVE = average variance extracted

5.5.6 Participants

The original sample contained 165 responses. Participants were eliminated if they did not finish the survey (n = 3), did not answer the attention check correctly (n = 11), or if their response pattern indicated that they did not recognize the experimental treatment (n = 30). The procedure to prepare the sample equaled the procedure applied during *Study 2* and *Study 3*. It led to a final sample of 121 responses used for further analyses. The final sample consisted of 55.4 percent male (n = 67) and 44.6 percent female (n = 54) participants. On average, they were 34.69 years old (SD = 10.163 years).

5.5.7 Results

5.5.7.1 Common-method Bias

To address the issue of CMV, the same a priori and statistical remedies as previously discussed were applied, including the presentation of the measured constructs on different pages of the questionnaire. In addition, the marker variable procedure was again applied, which was inserted into the model according to the CLC approach (Chin et al., 2013) to eliminate CMV effects. The scale introduced in *Study 2* to measure job satisfaction served as marker variable.

5.5.7.2 Treatment Check

Following *Study 2* and *Study 3*, as a first step, the sample was investigated in terms of the experimental treatment. As it was the case during the previous studies, some participants indicated that they had not recognized the experimental treatment (Sigall & Mills, 1998). Thus, subjects were removed if they rated the perceived emotions as extraordinarily positive while being assigned to the control condition or rated the perceived emotions as extraordinarily low in positivity while being assigned to the manipulated condition. This procedure led to the removal of 30 participants. Again, before the procedure was applied, a t-test comparing the groups was calculated already indicating significant differences ($M_{control}$ = 4.679, SD = 1.068, $M_{manipulation}$ = 5.252, SD = 1.211, 95% CI = −.940 to −.206, $t(147.147) = -3.083$, $p < .01$). From this, the conclusion was drawn that the manipulations achieved the intended effects and removing participants would not bias the results.

5.5.7.3 Path Model Estimation

The proposed model was estimated using SmartPLS v.3.3.3. Displaying positive emotions was operationalized as a dichotomous variable (no emotions = 0, positive emotions = 1). All other variables were specified as latent constructs. Table 5.18 presents the constructs and their corresponding items. Interaction terms of the independent and the moderating variables were calculated to test the moderating effects. Inference tests were calculated using a bootstrap procedure with 5,000 resamplings.

5.5.7.4 Measurement Model

The loadings extracted from SmartPLS revealed that all indicators loaded at least .723 on their corresponding construct. Therefore, the results indicated good reliability of the used indicators as they were all above the threshold of .70 (Hulland, 1999) (Table 5.18). Based on CR, all constructs furthermore demonstrated

Table 5.19 Discriminant Validity (Cross Loadings), Study 4

	PDE	CPA	EP	ES	EX	OE	NFI	EDB	EXPLC	EXPCB	Age	Gender
PDE	*1.000*	.245	.289	.238	−.027	.177	.076	−.093	−.002	.136	−.089	−.174
CPA_01	.133	*.731*	.468	.598	.054	.273	.093	−.014	.167	.043	.071	−.022
CPA_02	.278	*.787*	.453	.588	.163	.354	.297	−.076	.143	.109	−.014	−.037
CPA_03	.134	*.897*	.494	.603	.230	.288	.159	−.011	.186	.103	−.035	−.025
CPA_04	.200	*.767*	.442	.364	.219	.111	.055	.134	.020	.207	.000	−.016
CPA_05	.285	*.746*	.494	.454	.020	.134	.097	.111	.100	.180	.071	.085
CPA_06	.135	*.820*	.548	.526	.059	.253	.054	.087	.132	−.058	−.029	.059
EP_01	.234	.431	*.723*	.281	−.009	.038	−.074	.116	−.083	.098	.016	.089
EP_02	.252	.455	*.747*	.310	.038	.083	−.030	.029	−.146	.000	.070	.106
EP_03	.184	.467	*.761*	.634	−.062	.197	.043	.051	.157	−.016	−.058	.085
ES_01	.227	.607	.522	*.873*	.046	.333	.194	−.106	.167	−.001	.008	−.064
ES_02	.262	.495	.502	*.843*	−.028	.453	.266	−.048	.208	−.117	−.012	−.062
ES_03	.130	.558	.437	*.855*	.069	.325	.137	−.079	.131	−.006	−.041	−.053
ES_04	.248	.480	.512	*.770*	.032	.263	.081	.085	.005	−.007	.095	.027
ES_05	.128	.639	.528	*.814*	−.003	.186	.050	.037	.078	.011	−.086	.058
EX_01	−.007	.174	−.019	.056	*.965*	.245	.350	.141	.132	.232	.031	−.160
EX_02	−.067	.059	−.035	−.057	*.736*	.275	.224	.163	.144	.195	.086	−.085
OE_01	.090	.230	.106	.299	.386	*.848*	.294	.052	.209	.003	.093	−.025
OE_02	.209	.292	.172	.327	.072	*.822*	.354	.117	.247	.064	.146	.029

(continued)

Table 5.19 (continued)

	PDE	CPA	EP	ES	EX	OE	NFI	EDB	EXPLC	EXPCB	Age	Gender
NFI_01	.089	.205	.029	.185	.289	.319	*.902*	−.017	.130	.106	.045	−.237
NFI_02	.088	.133	.006	.155	.259	.388	*.866*	.087	.207	−.041	.040	−.168
NFI_03	.024	.104	−.069	.124	.377	.333	*.885*	.036	.202	.016	.120	−.098
EDB_01	−.126	.051	.091	−.013	.208	−.001	.048	*.764*	−.088	.157	−.069	.275
EDB_02	−.063	.022	.065	−.033	.116	.132	.021	*.952*	.042	.170	.121	.290
EXPLC	−.002	.166	.012	.145	.150	.272	.197	.000	*1.000*	.299	−.095	.054
EXPCB	.136	.118	.025	−.029	.246	.039	.040	.186	.299	*1.000*	−.043	−.102
Age	−.089	.012	−.004	−.012	.052	.142	.077	.067	−.095	−.043	*1.000*	.099
Gender	−.174	.006	.123	−.023	−.155	.001	−.194	.320	.054	−.102	.099	*1.000*

Notes: PDE = positive displayed emotions, CPA = customer positive affect, EP = empathy, ES = encounter satisfaction, EX = extraversion, OE = openness to experience, NFI = need for interaction, EDB = emotional decision behavior, EXPLC = experience with live chats, EXPCB = experience with chatbots acceptable reliability scores, as they were all equal to or above .788 (Table 5.18). This was estimated according to the suggested threshold of .70 (Bagozzi & Yi, 1988). Convergent validity was estimated by calculating the AVE. Table 5.18 shows that the AVEs in all cases were equal to or exceeded a value of .553 and thus were above the threshold of .50 (Bagozzi & Yi, 1988). Based on these scores, the conclusion was drawn that convergent validity was established.

As a final step, discriminant validity was evaluated. Consulting the cross loadings, no violation against discriminant validity was found, as all indicators loaded highest on their intended latent construct (Table 5.19). In the further course, again, the procedure suggested by Fornell and Larcker (1981) and the HTMT method (Henseler et al., 2015) were consulted. From Table 5.20 can be drawn that the square roots of the AVEs exceeded any of the correlations with the other variables of the model. Therefore, based on the Fornell-Larcker criterion, no violation against discriminant validity was found. Furthermore, applying the HTMT method, a maximum ratio of .804 indicated no violation against discriminant validity when considering the conservative threshold of .85 (Kline, 2011) (Table 5.21).

Table 5.20 Discriminant Validity (Fornell-Larcker Criterion), Study 4

	Latent construct	1	2	3	4	5	6	7
1	Customer positive affect							
2	Empathy	.608						
3	Encounter satisfaction	.673	.602					
4	Extraversion	.158	−.026	.027				
5	Openness to experience	.311	.165	.374	.281			
6	Need for interaction	.172	−.011	.176	.349	.387		
7	Emotional decision behavior	.035	.082	−.030	.164	.100	.033	
Square root of the AVE		.793	.744	.832	.858	.835	.885	.863

Notes: Scores indicate the correlation between the constructs

Table 5.21 Discriminant Validity (HTMT), Study 4

		1	2	3	4	5	6	7	8	9	10	11	12
1	Positive displayed emotions												
2	Customer positive affect	.261											
3	Empathy	.371	.804										
4	Encounter satisfaction	.254	.740	.720									
5	Extraversion	.051	.169	.075	.088								
6	Openness to experience	.239	.433	.244	.530	.463							
7	Need for interaction	.082	.191	.102	.198	.427	.562						
8	Emotional decision behavior	.130	.157	.140	.122	.272	.176	.078					
9	Experience with live chats	.002	.168	.214	.150	.189	.362	.218	.089				
10	Experience with chatbots	.136	.157	.063	.036	.293	.053	.066	.225	.299			
11	Age	.089	.049	.080	.062	.080	.190	.083	.131	.095	.043		
12	Gender	.174	.055	.155	.068	.168	.043	.204	.389	.054	.102	.099	

Table 5.22 Descriptives and Correlations, Study 4

		M/%	SD	1	2	3	4	5	6	7	8	9	10	11	12
1	Positive displayed emotions	50.4%	–												
2	Customer positive affect	5.173	1.057	.245											
3	Empathy	4.495	1.123	.289	.608										
4	Encounter satisfaction	5.687	1.211	.238	.673	.602									
5	Extraversion	4.532	1.522	−.027	.158	−.026	.027								
6	Openness to experience	5.222	1.267	.177	.311	.165	.374	.281							
7	Need for interaction	4.983	1.528	.076	.172	−.011	.177	.349	.386						
8	Emotional decision behavior	3.638	1.470	−.093	.035	.082	−.030	.164	.100	.033					
9	Experience with live chats	5.397	1.715	−.002	.166	.012	.145	.150	.272	.197	.000				

(continued)

Table 5.22 (continued)

		M/%	SD	1	2	3	4	5	6	7	8	9	10	11	12
10	Experience with chatbots	4.678	1.907	.136	.118	.025	−.029	.246	.039	.040	.186	.299			
11	Age	34.690	10.163	−.089	.012	−.004	−.012	.052	.142	.077	.067	−.095	−.043		
12	Gender (female)	44.6%	–	−.174	.006	.123	−.023	−.155	.001	−.194	.320	.054	−.102	.099	

Notes: Correlations equal to or above |.186| are statistically significant ($p < .05$, two-tailed).

5.5.7.5 Structural Model

The same scores as before were calculated to evaluate the performance of the structural model. R^2 showed, as expected, due to the additional explanatory construct empathy that the explained variance of encounter satisfaction indeed increased to .563. Displaying positive emotions explained around the same amount of variance of customer positive affect ($R^2 = .190$) and of empathy ($R^2 = .137$) (Figure 5.8). Predictive relevance could also be affirmed. This assumption was based on the extracted Q^2 values. Those were all equal to or above .013 and thus exceeded the underlying threshold of 0 (Chin, 1998; Henseler et al., 2009) (Figure 5.8). Concerning collinearity of the contained constructs, no violations against the threshold of 3 were found as all VIFs were either below or equal to 1.758 (Hair et al., 2019) (Table 5.23).

Table 5.23 Variance Inflation Factors, Study 4

		2	3	4
	Main effects			
1	Positive displayed emotions	1.125	1.167	1.226
2	Customer positive affect			1.758
3	Empathy			1.757
4	Encounter satisfaction			
	Moderating effects			
	Positive displayed emotions x extraversion	1.070		
	Positive displayed emotions x openness to experience		1.128	
	Controls			
	Extraversion	1.274		
	Openness to experience		1.370	
	Need for interaction	1.311	1.354	1.226
	Emotional decision behavior	1.296	1.254	1.249
	Experience with live chats	1.206	1.315	1.226
	Experience with chatbots	1.271	1.236	1.229
	Age	1.054	1.073	1.047
	Gender	1.335	1.338	1.312

5.5.7.6 Hypothesis Testing

In line with the previous results, the chatbot's display of positive emotions positively and significantly influenced customers' positive affect ($\beta = .224$, $p < .05$). Furthermore, the latter did not only positively impact encounter satisfaction ($\beta = .434$, $p < .001$) but *Study 4* could again deliver support for the mediation hypothesis (H2). This was because the indirect effect of positive displayed emotions through customer positive affect on encounter satisfaction was statistically significant (positive displayed emotions ➜ customer positive affect ➜ encounter satisfaction = .097, $p < .05$).

H3 hypothesized that empathy, the cognitive counterpart to primitive emotional contagion, acts as a second mediation path from positive displayed emotions on encounter satisfaction. Indeed, a positive effect of positive displayed emotions on empathy was present and significant ($\beta = .285$, $p < .01$). Furthermore, empathy positively influenced the satisfaction with the service encounter ($\beta = .332$, $p < .001$). Accordingly, the indirect effect of positive emotions through empathy on encounter satisfaction was assessed. The indirect effect of positive displayed emotions on encounter satisfaction with empathy acting as a mediator was statistically significant (positive displayed emotions ➜ empathy ➜ encounter satisfaction = .095, $p < .05$), thus delivering support for H3. As the direct path from positive displayed emotions to encounter satisfaction was statistically insignificant when the mediators were added to the model ($\beta = .002$, $p = .978$), the results delivered strong support for a full mediation through customer positive affect and empathy. The indirect effect through customer positive affect accounted for 51% of the total effect. The more conscious process through empathy accounted for the remaining 49%.

Supporting H4, the interaction term of positive displayed emotions and extraversion had a positive and significant effect on customer positive affect ($\beta = .248$, $p < .01$). This means that extraversion positively moderated the investigated effect. In contrast to the expectations of H5, the interaction term of positive displayed emotions and openness to experience did not have a significant effect on empathy ($\beta = .021$, $p = .842$). Hence, the effect of positive displayed emotions on empathy was not moderated by openness to experience. Therefore, H5 had to be rejected. Table 5.24 provides an overview of all path coefficients and their significances including all control variables.

Concerning the effect sizes, data indicated weak effects of positive displayed emotions on customer positive affect ($f^2 = .055$) and on empathy ($f^2 = .081$). In turn, customer positive affect had a moderate effect on encounter satisfaction ($f^2 = .245$), while empathy showed a weak effect on encounter satisfaction ($f^2 = .143$) (see Table 5.25 for an overview of all effect sizes).

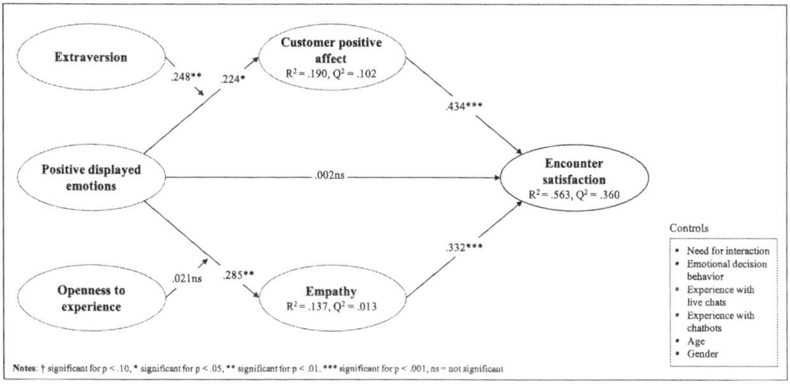

Figure 5.8 Results Structural Equation Modeling, Study 4

After consulting the path coefficients, their significances, and the effect sizes, the means for the investigated variables were calculated and compared across the two experimental groups. In line with the previous studies, subjects from the manipulated group reported significantly more positive affect ($M_{manipulation}$ = 5.430, SD = 1.092) than did those who were assigned to the control condition ($M_{control}$ = 4.913, SD = .960, 95% CI = −.887 to −.147, t(117.522) = −2.766, $p < .01$). Likewise, they reported to be significantly more satisfied with the service they received ($M_{manipulation}$ = 5.972, SD = 1.125) compared to the subjects of the control condition ($M_{control}$ = 5.398, SD = 1.237, 95% CI = −.999 to −.147, t(117.540) = −2.665, $p < .01$). Supporting the expectation that positive displayed emotions would help people to better empathize with the chatbot, subjects of the manipulated condition scored higher in their reported empathy ($M_{manipulation}$ = 4.816, SD = 1.128) than the subjects of the control condition ($M_{control}$ = 4.169, SD = 1.028, 95% CI = −1.035 to −.259, t(118.327) = −3.298, $p < .01$).

Table 5.24 Standardized Path Coefficients and Significances, Study 4

	Standardized coefficient	Standard deviation	T-statistics
Main effects			
Positive displayed emotions ➜ customer positive affect	.224	.093	2.424*

(continued)

Table 5.24 (continued)

	Standardized coefficient	Standard deviation	T-statistics
Positive displayed emotions ➜ empathy	.285	.099	2.889**
Positive displayed emotions ➜ encounter satisfaction	.002	.061	.027
Customer positive affect ➜ encounter satisfaction	.434	.083	5.234***
Empathy ➜ encounter satisfaction	.332	.092	3.625***
Indirect effects			
Positive displayed emotions ➜ customer positive affect ➜ encounter satisfaction	.097	.046	2.100*
Positive displayed emotions ➜ empathy ➜ encounter satisfaction	.095	.040	2.383*
Moderating effects			
Positive displayed emotions x extraversion ➜ customer positive affect	.248	.092	2.701**
Positive displayed emotions x openness to experience ➜ empathy	.021	.106	.200
Control paths			
Extraversion ➜ customer positive affect	.100	.098	1.016
Openness to experience ➜ empathy	.142	.119	1.199
Need for interaction ➜ customer positive affect	.058	.100	.575
Need for interaction ➜empathy	–.074	.114	.647
Need for interaction ➜ encounter satisfaction	.044	.068	.653
Emotional decision behavior ➜ customer positive affect	–.009	.114	.081
Emotional decision behavior ➜empathy	.064	.130	.495
Emotional decision behavior ➜ encounter satisfaction	–.009	.082	.114
Experience with live chats ➜ customer positive affect	.125	.113	1.103
Experience with live chats ➜empathy	–.028	.138	.206

(continued)

Table 5.24 (continued)

	Standardized coefficient	Standard deviation	T-statistics
Experience with live chats ➜ encounter satisfaction	.084	.079	1.071
Experience with chatbots ➜ customer positive affect	.033	.100	.327
Experience with chatbots ➜ empathy	−.011	.110	.101
Experience with chatbots ➜ encounter satisfaction	−.126	.073	1.725†
Age ➜ customer positive affect	.048	.086	.561
Age ➜empathy	−.010	.097	.103
Age ➜ encounter satisfaction	−.002	.058	.037
Gender (female) ➜ customer positive affect	.091	.101	.909
Gender (female) ➜ empathy	.144	.109	1.317
Gender (female) ➜ encounter satisfaction	−.072	.068	1.049

Notes: † significant for $p < .10$, * significant for $p < .05$, ** significant for $p < .01$, *** significant for $p < .001$.

Table 5.25 Effect Sizes, Study 4

		2	3	4
	Main effects			
1	Positive displayed emotions	.055	.081	.000
2	Customer positive affect			.245
3	Empathy			.143
4	Encounter satisfaction			
	Moderating effects			
	Positive displayed emotions x extraversion	.071		
	Positive displayed emotions x openness to experience		.000	
	Controls			
	Extraversion	.010		

(continued)

Table 5.25 (continued)

		2	3	4
	Openness to Experience		.017	
	Need for interaction	.003	.005	.004
	Emotional decision behavior	.000	.004	.000
	Experience with live chats	.016	.001	.013
	Experience with chatbots	.001	.000	.030
	Age	.003	.000	.000
	Gender (female)	.008	.018	.009

5.5.8 Discussion

Study 4 was designed to pursue two objectives. The first one was to examine the role of empathy as a second mediator in the relationship of positive displayed emotions and encounter satisfaction. The second concerned extraversion and openness to experience as moderators of the investigated affective responses by customers. Empathy (i.e., the situational ability of being able to put yourself into the emotional state of the chatbot) was expected to represent the second parallel mediation path that is, in contrast to emotional contagion, more strongly characterized by consciousness. The results were able to show that empathy plays a mediating role in the relationship between positive displayed emotions and encounter satisfaction. With these results, the present thesis is one of the first to examine empathy from this perspective in the context of a service encounter. In this respect, the results show that the display of positive emotions leads to customers being more easily able to put themselves in the emotional state of a chatbot. This in turn leads to customers evaluating the experienced service significantly better. With the results, the thesis provides additional support for what was introduced as anthropomorphism (Epley et al., 2007). In contrast to the CASA paradigm (Reeves & Nass, 1996), it is based on a more conscious level, just like empathy is compared to emotional contagion. In sum, one can draw the conclusion that through the typical human behavior (i.e., displaying positive emotions) on the one hand unconscious social responses are triggered in customers. On the other hand, this behavior also leads to customers anthropomorphizing the chatbot more strongly. Regarding the influencing role of empathy on encounter satisfaction, the study delivers comparable results to the scarce literature on empathy's effect during service encounters but investigated from a customer's perspective (e.g., Gremler & Gwinner, 2000; Wieseke et al., 2012).

Empathy toward an employee in such situations serves as an evaluation of abilities (Gremler & Gwinner, 2000). These findings provide additional support for the influence social components have during chatbot-based service encounters. Namely, the affective delivery acts as an indicator of human likeness that triggers anthropomorphic behavior toward the chatbot. Besides the indirect effect through empathy, the indirect through customer positive affect remained significant. From this can be concluded, that empathy and customer positive affect are able to fully explain customers' affective responses to a chatbot's positive displayed emotions. The second objective of the study was to examine the moderating roles of extraversion and openness to experience on customers' affective responses. The results clearly showed that in customer segments characterized by high levels of extraversion, the unconscious effects resulting from positive displayed emotions of a chatbot are stronger. Contrary to expectations, openness to experience had no moderating influence on the ability to empathize with the chatbot.

What is important to mention, neither a physical representation is necessary for an anthropomorphic behavior (i.e., the attribution of human-like characteristics) nor is any graphical representation. Instead, contrary to previous research on robots that, due to their physical appearance, more easily produce anthropomorphic behavior (Duffy, 2003), even only virtually existing entities can trigger such behavioral patterns only based on their behavior. However, with advancing technical capabilities, it has become increasingly popular in recent years to display chatbots graphically as well. Since it has not yet been clarified whether and to what extent this (i.e., graphically representing chatbots) impacts subjects' emotional reactions to the chatbot's affective delivery, the following *Study 5* focuses on specifically this question.

5.6 Study 5: Effect of Avatars on Anthropomorphic Behavior Toward Chatbots[5]

5.6.1 Conceptual Development

In their research on conversational agents, Heerink et al. (2010) pointed out that the CASA paradigm also applies to interactions between chatbots and humans. However, it should be emphasized here that their research featured an agent with a graphical representation. This stream of research focuses on the fact that it is

[5] Parts of *Study 5* have been presented at the 7th Rostock Conference on Service Research.

primarily the graphical representation that leads to the triggering of social behavior toward a conversational agent by a human (e.g., Corti & Gillespie, 2016; Lee et al., 2010; von der Pütten et al., 2010). Up to this point, the results of the present thesis showed that, on both a subconscious and a conscious level, a graphical representation of the conversational agent is not necessary to trigger affective responses, such as emotional contagion and empathy of humans toward chatbots. Instead, the results show that it strongly depends on the chatbot's behavioral characteristics (i.e., the expression of positive emotions).

In human-to-chatbot interactions, the issue of social behavior, however, is not only relevant in terms of social behavior toward chatbots. Instead, this question is increasingly also considered explicitly from the perspective of under which conditions a chatbot is perceived as a social actor (e.g., Heerink et al., 2010; van Doorn et al., 2017). In their experiment, Riek, Rabinowitch, Chakrabarti, and Robinson (2009a) manipulated human likeness through the graphical representation in a video. This is similar to the approach seen in research on conversational agents. For example, Lin et al. (2021) report that the display of human-like avatars offer the potential that the interaction with a conversational agent is more strongly perceived as being with another social being. Bente et al. (2008) note that this influence of an avatar is due to its ability to transport nonverbal and relational information. The representation of avatars can range from relatively simple static graphics (e.g., Vassos et al., 2016) to complex three-dimensional versions (e.g., Nunamaker, Derrick, Elkins, Burgoon, & Patton, 2011).

In line with the previous results regarding the creation of the feeling of being with another social being, scholars have found that this feeling can also be created by a chatbot through the use of informal language, a human name, and the general reference of communication to human communication (greetings and goodbyes) (Araujo, 2018). The latter, in particular, is understandable in so far as the classical face-to-face communication is, so to speak, the ideal reference for the conception of computer-mediated communication (Rüggenberg, 2007).

Overall, these results imply that typically human behaviors increase the feeling of being in the presence of a social being. At the same time, perceived human likeness is shown to play a central role in explaining social reactions of humans toward conversational agents and other artificial entities. Since nowadays, the communication of emojis in text-based communication belongs to a central property of human-to-human communication, it becomes apparent that the present thesis has addressed up to this point a typically human behavior (Felbo et al., 2017). For the further course, the assumption is made that human attributes act as a kind of cue to trigger social responses and with more human-like cues being shown, the stronger the anthropomorphizing effects appear (Blut et al.,

2021). In contrast, these effects are expected to be weaker when a computer-like avatar is shown, which emphasizes the technical nature of the chatbot. While several human-like cues are shown in the human-like avatar condition, and it is assumed that they mutually reinforce each other, only the behavior (i.e., the display of positive emotions) is the only human-like cue in the computer-like avatar condition. Thus, as a supplement to the already existing behavioral attributes, the graphical representation of the chatbot is specifically addressed in the further course. Summarizing the aforementioned arguments, the following is hypothesized (Figure 5.9):

> **H6**: The human likeness of a chatbot's graphical representation moderates the positive effect of a chatbot's positive displayed emotion on customer positive affect, such that the effect will be stronger when a human-like avatar is shown compared to a computer-like avatar.
> **H7**: The human likeness of a chatbot's graphical representation moderates the positive effect of a chatbot's positive displayed emotion on empathy, such that the effect will be stronger when a human-like avatar is shown compared to a computer-like avatar.

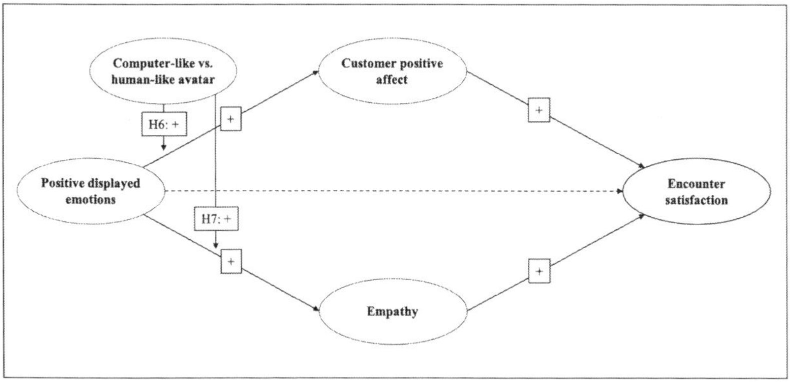

Figure 5.9 Proposed Research Model and Hypothesis, Study 5

5.6.2 Research Objectives and Setting

To follow the objective of *Study 5,* (i.e., to shed more light on the effect of graphical representations of chatbots on the already known effects caused by displaying positive emotions) the already known scenario around booking a hotel room served as the basis. This scenario was a good starting point to carry out the adaptations required for the following study (see Section 5.6.3).

5.6.3 Experimental Design

To assess the impact of using an avatar, especially on empathy, a 2×2 between-subjects design was used manipulating, on the one hand, the chatbot's emotional behavior (no emotions vs. positive emotions) and, on the other hand, the chatbot's graphical representation (computer-like avatar vs. human-like avatar). As the series of previously conducted studies showed that using videos as stimuli delivers reliable results, the stimuli of *Study 2* and *Study 4* were used and adapted. Before the adaption, the top area of the simulated display only showed information such as the chatbot's name and an indication whether the chatbot was currently typing a message. For *Study 5*, this area was supplemented by a picture of the corresponding avatar. A larger version of the avatar was displayed during the already described opening credits to strengthen the manipulation (Figure 5.10). As it was the case for the previous studies, displaying positive emotions again combined the 😀-emoji with added witty statements. As the overall procedure was intended to remain unchanged, settings for the videos equaled the previous ones. Thus, the videos again had a size of 1920 pixels (height) by 1080 pixels (width) and were created using Adobe After Effects v16.1.1 for MacOS.

For the identification, if the avatars indicated high or low human likeness as they were supposed to do, a pre-test was conducted. This test that used the four avatars depicted in Figure 5.11 led to the identification of two avatars (i.e., a computer-like and a human-like avatar) showing a clear and significant difference regarding the attributed human likeness. Right from the beginning, the computer-like avatar was intended to be represented by a picture highlighting the chatbot's technical and computer-based attributes. For this reason, a picture showing a piece of source code [1] was chosen. The pre-test supported the assumption this would adequately represent an avatar with strong computer-like characteristics.

The conceptualization of the human-like avatar proved to be somewhat more complex. For the pre-studies, a drawing of a robot-like upper body with a head [2], a robot with more human features being shown such as a nose [3], and a female upper body with a comparably high level of details [4] were prepared. The expectations concerning the discrimination between the avatars were a stepwise adding of human-like features, such as a head and a face. While those features were simply displayed as silhouettes for the first robot [2], they were shown with a higher level of details for the second robot [3] and ultimately human-like in the third avatar [4]. A one-way between-subjects ANOVA calculated following the pre-test showed no significant differences between the two robot avatars and the computer-like avatar showing source code. However, the female upper body was perceived as significantly more human-like than the computer-like source code. To have a clear and strong manipulation, the decision was made to use the two avatars (computer-like and human-like) for the further course of the study.

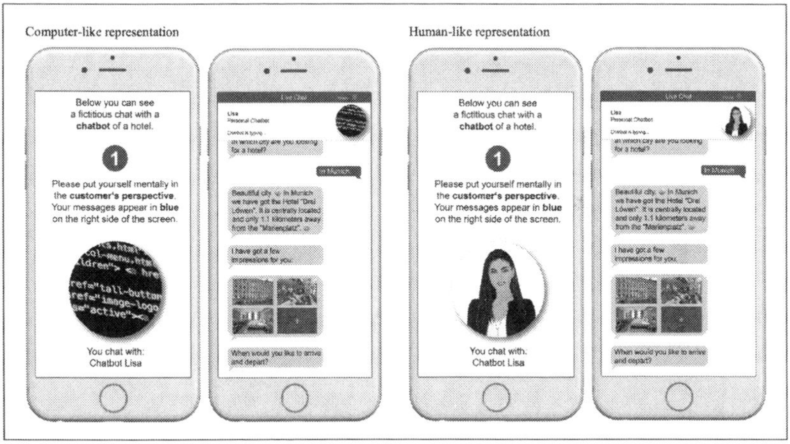

Figure 5.10 Graphical Manipulation of the Chatbot's Human Likeness, Study 5. (**Notes:** Chat protocols have been translated from German for this figure.)

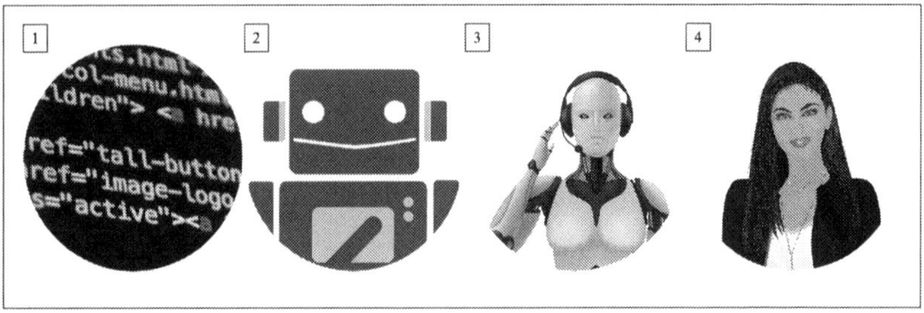

Figure 5.11 Tested Avatars to Graphically Manipulate Human Likeness, Study 5

5.6.4 Procedure

The procedure followed the approaches in the previously conducted and presented studies. As for the use of video-based stimuli, the previously used ones were taken and the avatar added. Based on the decision to use a web-based experiment to gather the data, the decision was made to once more rely on Clickworker as the source for participants. The same remedies as before were applied to ensure data quality collected from quasi-professional survey takers (e.g., monetary compensation and attention checks). The structure of the used questionnaire corresponded to previous investigations, as did the used measurement instruments.

5.6.5 Participants

Initially, 268 subjects participated in the web-based experiment. However, as in the three reported studies that relied on the same procedure, some participants had to be excluded for several reasons. Subjects were either removed from the sample if their responses were incomplete (n = 6), they failed to answer the attention check correctly (n = 25), or they appeared not to have recognized the experimental treatment (n = 46). This left a final sample of 191 subjects. Of those, 52.4 percent were male (n = 100) and 47.1 percent were female (n = 90). One participant indicated being neither male nor female. On average, the participants were 34.06 years old (SD = 12.306 years).

5.6.6 Results

5.6.6.1 Common-method Bias

CMV was addressed relying on the same a priori and statistical remedies applied during *Study 2, Study 3,* and *Study 4*. This included, for example, measuring constructs on different pages of the questionnaire. Additionally, the marker variable approach (Chin et al., 2013) was applied as a statistical remedy to eliminate CMV effects. The job satisfaction scale (Bowling & Hammond, 2008) introduced in *Study 2* served as marker variable.

5.6.6.2 Treatment Check

Consistent with the previous studies, a treatment check was performed for the presented *Study 5* before excluding participants based on their response to the experimental treatment. Again, the aim was to identify subjects whose response behavior indicated that they had not recognized the experimental treatment (Sigall & Mills, 1998). The procedure was identical to that of the previous studies and was based on the perceived positivity of the displayed emotions. Initially, an ANOVA was calculated to evaluate the overall manipulation. The results showed a significant effect across the four experimental conditions ($F(3, 233) = 5.267$, $p < .01$). The Levene's test that homogeneity of variances being established ($p = .649$). Thus, LSD post hoc tests were used to evaluate the means of positivity within the two avatar conditions. Based on this procedure, the conclusion could be drawn that participants in the computer-like avatar condition rated the perceived positivity of the displayed emotions as significantly higher when exposed to the manipulated positive emotions compared to the control condition ($M_{manipulation} = 5.328$, $SD = 1.072$, $M_{control} = 4.736$, $SD = 1.307$, $p < .01$). The same results could be found investigating the differences across the two human-like conditions ($M_{manipulation} = 5.119$, $SD = 1.267$, $M_{control} = 4.530$, $SD = 1.203$, $p < .01$). This indicated that the overall manipulation was successful. However, as previously encountered, several participants' answers indicated that they had not recognized the experimental treatment. Applying the same procedure as during the previous studies, a total of 46 subjects was removed from the data set. An additional treatment check for human likeness was not performed. The following reason was decisive. During the pre-test, only the avatars were shown to the participants and their human likeness was considered in isolation. In this respect, it did not seem feasible to measure the isolated human likeness of the avatar again in the main study. It is crucial to highlight that the human likeness of the respective avatars was considered and not the human likeness of the chatbot as a whole.

5.6.6.3 Path Model Estimation

The proposed model was again calculated using SmartPLS v3.3.3. For inferential calculations, a bootstrapping procedure with 5,000 resamplings was used. The scaling of the experimental variable was dichotomous (no emotions = 0, positive emotions = 1). To further test the hypotheses, multigroup analysis (MGA) was performed, for which the chatbot's graphical representation (computer-like vs. human-like) was used as the grouping variable. This procedure aims to analyze differences concerning the model between two groups. An advantage in this context is, although the specific assumption with regard to the influence of an avatar, an MGA opens up the possibility to investigate the influence of different avatars on the entire model in comparison to an isolated consideration of individual paths (Hair et al., 2014a). In addition to testing the specific assumptions of the hypotheses, this also allowed for an exploratory investigation of the remaining paths. To test for differences between models, SmartPLS offers several parametric and nonparametric procedures. In the present case, the nonparametric permutation procedure with 5,000 resamplings was applied, which is a comparatively conservative methodology to compare paths between models in a MGA (Chin & Dibbern, 2010; Hair, Sarstedt, Ringle, & Gudergan, 2018b; Matthews, 2017). For this reason, statistical differences were assessed on a 90% confidence interval.

5.6.6.4 Measurement Model and Measurement Invariance

The basis for the MGA is the review of both measurement models (i.e., a separate examination for both groups). This review followed the procedure applied during the previous studies. This means, the proposed procedure of Hair et al. (2012) is followed again. Starting with the indicator reliability, the factor loadings surpassed in all cases a value of .717 and were thus above the threshold of .70 (Hulland, 1999) (Table 5.26). The CR of all constructs exceeded the underlying threshold of .70 (Bagozzi & Yi, 1988) as all values were either equal to or above .887 (Table 5.29). Therefore, internally consistent results could be confirmed. This was also valid for convergent validity. The AVE was at least .688 and thus was in all cases above the suggested threshold of .50 (Bagozzi & Yi, 1988). Discriminant validity was again assessed, consulting the cross loadings. The cross loadings indicated no violation, as in both groups the indicators loaded highest on the intended constructs (Table 5.27). This applied also for the Fornell-Larcker criterion as well as the HTMT method. The square root of the AVEs exceeded the correlations with every other construct in all cases (Table 5.28). The HTMT method showed that the maximum ratio was .666 indicating no violation against discriminant validity as this ratio did not exceed the conservative threshold of .85

(Kline, 2011) (Table 5.29). These results indicated no violations against any of the underlying quality criteria within the two investigated subsamples.

In addition to examining the measurement models in isolation, however, it is necessary to inspect measurement invariance. This evaluation step aims at checking the two groups formed for the MGA for possible differences regarding the respective measurement models (Horn & McArdle, 1992). This is important to ensure that differences between groups are not due to differences in the measurement models. To check measurement invariance, SmartPLS relies on the so-called measurement invariance of composite models (MICOM) procedure, which is based on three steps to evaluate possible differences between subsamples (Henseler, Ringle, & Sarstedt, 2016). In a first step, the so-called configural invariance is considered. This is a qualitative evaluation step, aiming to check whether the same procedure was used to prepare the two subsamples and the composition of the indicators was identical. In a second step, the results of the permutation test were used to evaluate compositional invariance. In this respect, invariance can be assumed, if the correlation of a composite between the two groups is statistically not different from 1 (Henseler et al., 2016). If compositional and configural invariance can be confirmed, Henseler et al. (2016) consider this as a situation in which partial invariance is present. This would still be sufficient to continue with the analysis, however, an investigation with the pooled data would not be recommended. In a third and final step, the mean values and variances are consulted (Henseler et al., 2016). For both accounts, that measurement invariance can be assumed if the original differences values fall within the 95% confidence interval. If this criterion applies to both, Henseler et al. (2016) speak of full invariance. If it only applies to one, partial invariance could still be assumed.

In the present case, configural invariance could be approved since the preparatory work (e.g., the treatment check) was carried out for the pooled data set. Furthermore, the indicators used to form the composites in both groups were also identical. According to the permutation results, the conclusion could furthermore be drawn that the correlations were statistically not different from 1, indicating compositional invariance. Lastly, the MICOM procedure revealed a significantly differing result concerning empathy and the related means. In contrast, the results showed that equal variances were given. Therefore, partial invariance could be confirmed, meaning a multigroup analysis was possible. Table 5.30 and Table 5.31 summarize the results of the three evaluation steps.

Table 5.26 Items and Indicator Loadings for Latent Constructs, Study 5

Construct	Computer-like avatar	Human-like avatar
Main constructs		
Customer positive affect		
I feel strong.	.729	.754
I feel elated.	.862	.815
I feel excited.	.894	.853
I feel enthusiastic.	.894	.858
I feel active.	.887	.867
I feel peppy.	.877	.823
Empathy		
I was very close to the chatbot's emotions.	.883	.925
I felt compassionate toward the chatbot.	.884	.905
I felt soft-hearted toward the chatbot.	.782	.734
Encounter satisfaction		
I am delighted by the experienced service.	.854	.884
The service helped me to solve my concern.	.870	.717
It was a good service experience.	.917	.936
I am satisfied with the experienced service.	.917	.939
I really liked the service experience.	.845	.885
Controls		
Need for interaction		
I feel more relaxed when I have personal contact with service employees.	.912	.854
I rather like to communicate with people when services are provided.	.935	.885
I prefer personal contact for asking and answering my questions.	.858	.916
Emotional decision behavior		
In general, in decision making, I tend to be guided by my rational side/my emotional side.	.958	.922
In general, in decision making, I tend to be guided by my head/my heart.	.899	.916

Table 5.27 Discriminant Validity (Cross Loadings), Study 5

	PDE	CPA	EP	ES	NFI	EDB	EXPLC	EXPCB	Age	Gender
Computer-like avatar										
PDE	*1.000*	.027	.188	.233	−.002	.257	.087	.111	−.031	.191
CPA_01	−.012	*.729*	.361	.484	.066	−.171	.017	.176	−.015	−.077
CPA_02	−.033	*.862*	.419	.364	−.043	.016	.005	.134	−.074	.071
CPA_03	−.055	*.894*	.480	.516	−.012	.020	−.136	.049	−.167	.065
CPA_04	.030	*.894*	.452	.436	−.032	.154	−.209	.067	−.141	.077
CPA_05	.122	*.887*	.466	.428	.011	.167	−.121	.062	−.001	.031
CPA_06	.108	*.877*	.487	.385	−.039	.135	−.166	−.015	.011	−.051
EP_01	.065	.450	*.883*	.243	−.082	.094	−.205	.023	−.028	−.058
EP_02	.162	.487	*.884*	.259	−.049	.179	−.110	.141	−.037	−.056
EP_03	.233	.383	*.782*	.318	−.212	.071	.071	.243	−.154	.062
ES_01	.180	.381	.241	*.854*	−.161	.004	−.102	−.026	−.047	.192
ES_02	.230	.424	.186	*.870*	−.058	−.017	.112	.075	−.037	.237
ES_03	.228	.481	.213	*.917*	−.137	.061	.031	.098	−.008	.269
ES_04	.174	.481	.307	*.917*	−.079	.010	−.013	.146	−.046	.192
ES_05	.211	.490	.477	*.845*	−.139	.118	−.152	.066	−.038	.183
NFI_01	−.031	−.022	−.174	−.146	*.912*	−.236	−.206	−.079	.211	−.106
NFI_02	.012	−.072	−.144	−.153	*.935*	−.162	−.138	−.036	.263	−.109
NFI_03	.019	.104	−.035	−.033	*.858*	−.031	−.066	.016	.183	−.167

(continued)

Table 5.27 (continued)

	PDE	CPA	EP	ES	NFI	EDB	EXPLC	EXPCB	Age	Gender
EDB_01	.241	.077	.135	.074	−.206	*.958*	−.151	−.178	−.042	.134
EDB_02	.239	.020	.113	−.011	−.086	*.899*	−.020	−.037	.094	.281
EXPLC	.087	−.121	−.086	−.028	−.159	−.106	*1.000*	.666	.034	.026
EXPCB	.111	.093	.169	.084	−.042	−.130	.666	*1.000*	−.048	−.008
Age	−.031	−.081	−.091	−.039	.246	.012	.034	−.048	*1.000*	−.140
Gender	.191	.024	−.016	.245	−.136	.206	.026	−.008	−.140	*1.000*
Human-like avatar										
PDE	*1.000*	.300	.296	.222	.059	.059	.315	.145	−.063	−.043
CPA_01	.160	*.754*	.165	.505	−.058	.226	.226	.047	.051	.111
CPA_02	.143	*.815*	.312	.440	−.233	.272	.194	.152	−.028	.185
CPA_03	.178	*.853*	.447	.562	−.196	.255	.168	.190	−.114	−.002
CPA_04	.313	*.858*	.373	.428	−.213	.206	.249	.186	−.185	.142
CPA_05	.319	*.867*	.427	.458	−.211	.150	.190	.124	.007	.176
CPA_06	.364	*.823*	.398	.527	−.183	.263	.165	.072	.135	.170
EP_01	.344	.383	*.925*	.425	−.216	.237	.054	.089	.056	.059
EP_02	.283	.356	*.905*	.298	−.263	.227	.030	.113	.024	.056
EP_03	.104	.371	*.734*	.394	−.224	.081	.031	−.067	.123	.019
ES_01	.174	.572	.392	*.884*	−.115	.124	.295	.115	.005	.066
ES_02	.252	.426	.198	*.717*	.088	−.021	.220	.077	−.056	.070

(continued)

Table 5.27 (continued)

	PDE	CPA	EP	ES	NFI	EDB	EXPLC	EXPCB	Age	Gender
ES_03	.160	.497	.333	*.936*	−.117	.221	.176	.186	−.039	.027
ES_04	.189	.503	.390	*.939*	−.125	.218	.122	.098	−.025	.078
ES_05	.210	.556	.517	*.885*	−.015	.254	.209	.210	−.121	.091
NFI_01	−.001	−.279	−.245	−.083	*.854*	−.076	.034	.002	−.022	−.025
NFI_02	.099	−.153	−.228	−.047	*.885*	−.033	.048	−.008	.044	−.027
NFI_03	.069	−.139	−.242	−.053	*.916*	−.038	.014	.075	.061	−.064
EDB_01	.078	.283	.166	.199	−.078	*.922*	−.113	.006	−.034	.187
EDB_02	.030	.225	.239	.156	−.027	*.916*	−.132	−.030	−.017	.189
EXPLC	.315	.239	.046	.235	.036	−.133	*1.000*	.558	−.157	−.037
EXPCB	.145	.154	.061	.162	.026	−.013	.558	*1.000*	−.159	−.022
Age	−.063	−.024	.075	−.056	.029	−.028	−.157	−.159	*1.000*	−.013
Gender	−.043	.157	.054	.077	−.043	.204	−.037	−.022	−.013	*1.000*

Notes: PDE = positive displayed emotions, CPA = customer positive affect, EP = empathy, ES = encounter satisfaction, NFI = need for interaction, EDB = emotional decision behavior, EXPLC = experience with live chats, EXPCB = experience with chatbots

Table 5.28 Discriminant Validity (Fornell-Larcker Criterion), Study 5

	Latent construct	1	2	3	4	5
	Computer-like avatar					
1	Customer positive affect					
2	Empathy	.518				
3	Encounter satisfaction	.515	.325			
4	Need for interaction	−.007	−.139	−.131		
5	Emotional decision behavior	.058	.135	.043	−.170	
	Square root of the AVE	.859	.851	.881	.902	.929
	Human-like avatar					
1	Customer positive affect					
2	Encounter satisfaction	.429				
3	Empathy	.589	.433			
4	Need for interaction	−.221	−.270	−.070		
5	Emotional decision behavior	.277	.219	.193	−.057	
	Square root of the AVE	.829	.859	.876	.885	.919

Notes: Scores indicate the correlation between the constructs

Table 5.29 Discriminant Validity (HTMT), Study 5

					Discriminant validity (HTMT)									
		CA	CR	AVE	1	2	3	4	5	6	7	8	9	10
	Computer-like avatar													
1	Positive displayed emotions	–	–	–										
2	Customer positive affect	.928	.944	.738	.073									
3	Empathy	.808	.887	.724	.201	.598								
4	Encounter satisfaction	.928	.945	.776	.241	.545	.369							
5	Need for interaction	.886	.929	.814	.024	.090	.163	.144						
6	Emotional decision behavior	.848	.926	.863	.280	.142	.160	.077	.182					
7	Experience with live chats	–	–	–	.087	.132	.169	.097	.161	.100				
8	Experience with chatbots	–	–	–	.111	.101	.177	.097	.051	.125	.666			
9	Age	–	–	–	.031	.082	.096	.041	.258	.079	.034	.048		
10	Gender	–	–	–	.191	.075	.076	.253	.150	.242	.026	.008	.140	
	Human-like avatar													
1	Positive displayed emotions	–	–	–										
2	Customer positive affect	.909	.929	.688	.312									
3	Empathy	.818	.893	.737	.315	.499								
4	Encounter satisfaction	.922	.942	.767	.235	.637	.486							

(continued)

Table 5.29 (continued)

					Discriminant validity (HTMT)									
5	Need for interaction	.862	.916	.784	.069	.247	.324	.118						
6	Emotional decision behavior	.816	.916	.844	.065	.320	.261	.221	.065					
7	Experience with live chats	–	–	–	.315	.252	.049	.244	.039	.148				
8	Experience with chatbots	–	–	–	.145	.163	.116	.163	.034	.022	.558			
9	Age	–	–	–	.063	.110	.088	.059	.051	.031	.157	.159		
10	Gender	–	–	–	.043	.166	.058	.079	.047	.226	.037	.022	.013	

Notes: CA = Cronbach's alpha, CR = composite reliability, AVE = average variance extracted

Table 5.30 MICOM Steps 1 and 2, Study 5

	Step 1	Step 2		
Construct	Configural invariance	C = 1	5% quantile	Compositional invariance
CPA	Yes	.999	.993	Yes
EP	Yes	.996	.987	Yes
ES	Yes	.999	.998	Yes
NFI	Yes	.999	.968	Yes
EDB	Yes	.997	.952	Yes

Notes: CPA = customer positive affect, EP = empathy, ES = encounter satisfaction, NFI = need for interaction, EDB = emotional decision behavior

Table 5.31 MICOM Step 3, Study 5

	Step 3a		Step 3b		
Construct	Differences CI mean value	Equal mean values	Differences CI variances	Equal variances	Measurement invariance
CPA	−.028 [−.278 to .277]	Yes	.393 [−.489 to .492]	Yes	Full
EP	.312 [−.279 to .286]	No	.164 [−.351 to .352]	Yes	Partial
ES	−.032[−.280 to .285]	Yes	−.090 [−.611 to .605]	Yes	Full
NFI	−.095[−.278 to .274]	Yes	.077 [−.398 to .419]	Yes	Full
EDB	.150 [−.291 to .279]	Yes	.249 [−.336 to .328]	Yes	Full

Notes: CPA = customer positive affect, EP = empathy, ES = encounter satisfaction, NFI = need for interaction, EDB = emotional decision behavior

Table 5.32 Descriptives and Correlations, Study 5

		M/%	SD	1	2	3	4	5	6	7	8	9	10
Computer-like avatar													
1	Positive displayed emotions	48.9%	–										
2	Customer positive affect	5.171	1.210	.027									
3	Empathy	4.567	1.314	.188	.518								
4	Encounter satisfaction	5.937	.997	.233	.515	.325							
5	Need for interaction	5.267	1.401	−.002	−.007	−.139	−.131						
6	Emotional decision behavior	3.674	1.580	.257	.058	.135	.043	−.170					
7	Experience with live chats	5.543	1.727	.087	−.121	−.086	−.028	−.159	−.106				
8	Experience with chatbots	4.787	1.690	.111	.093	.169	.084	−.042	−.130	.666			
9	Age	34.340	12.368	−.031	−.081	−.091	−.039	.246	.012	.034	−.048		
10	Gender (female)	46.8%	-	.191	.024	−.016	.245	−.136	.206	.026	−.008	−.140	
Human-like avatar													
1	Positive displayed emotions	48.5%	-										

(continued)

Table 5.32 (continued)

		M/%	SD	1	2	3	4	5	6	7	8	9	10
2	Customer positive affect	5.154	1.002	.300									
3	Empathy	4.056	1.253	.296	.429								
4	Encounter satisfaction	5.957	1.047	.222	.589	.433							
5	Need for interaction	5.389	1.353	.059	−.221	−.271	−.070						
6	Emotional decision behavior	3.448	1.390	.059	.277	.219	.193	−.057					
7	Experience with live chats	5.660	1.779	.315	.239	.046	.235	.036	−.133				
8	Experience with chatbots	4.536	1.832	.145	.154	.061	.162	.026	−.013	.558			
9	Age	33.790	12.303	−.063	−.024	.075	−.056	.029	−.028	−.157	−.159		
10	Gender (female)	47.4%	-	−.043	.157	.054	.077	−.044	.205	−.037	−.022	−.013	

Notes: Correlations equal to or above |.205| are statistically significant ($p < .05$, two-tailed).

5.6.6.5 Structural Model

After checking whether there are differences in the underlying measurement models, the next step is to evaluate the structural models. Based on the structural model evaluation, the results across the two avatar groups clearly indicated the R^2 scores of the endogenous variables were higher in the human-like group (i.e., the group where a human-like avatar was shown). The average R^2 score in the human-like group was .313, whereas it was only .215 in the computer-like group (i.e., the group where a computer-like avatar was shown). Taking a closer look, the results revealed mainly differences in customer positive affect being responsible for the large differences. Displaying positive emotions was in the human-like group to a larger extent able to explain customer positive affect ($R^2 = .295$) than it could do in the computer-like group ($R^2 = .090$). The R^2 scores for the remaining endogenous variables were also higher in the human-like group, however, the differences were smaller (Figure 5.12 and Figure 5.13). For both models, predictive relevance could be confirmed as all Q^2 values were equal to or above .041 thus exceeding 0 (Chin, 1998; Henseler et al., 2009) (Figure 5.12 and Figure 5.13). All VIFs were equal to or below 2.125, thus indicating that no collinearity between the constructs existed as they did not exceed the threshold of 3 (Hair et al., 2019) (Table 5.33).

Table 5.33 Variance Inflation Factors, Study 5

		2	3	4
	Computer-like avatar			
1	Positive displayed emotions	1.128	1.128	1.172
2	Customer positive affect			1.426
3	Empathy			1.579
4	Encounter satisfaction			
	Need for interaction	1.191	1.191	1.235
	Emotional decision behavior	1.189	1.189	1.198
	Experience with live chats	1.907	1.907	2.118
	Experience with chatbots	1.916	1.917	2.125
	Age	1.160	1.160	1.165
	Gender	1.105	1.105	1.117

(continued)

Table 5.33 (continued)

		2	3	4
Human-like avatar				
1	Positive displayed emotions	1.146	1.194	1.297
2	Customer positive affect			1.562
3	Empathy			1.411
4	Encounter satisfaction			
	Need for interaction	1.025	1.016	1.169
	Emotional decision behavior	1.090	1.090	1.191
	Experience with live chats	1.655	1.657	1.742
	Experience with chatbots	1.517	1.513	1.524
	Age	1.070	1.039	1.088
	Gender	1.048	1.049	1.068

5.6.6.6 Hypothesis Testing

H6 proposed displaying a human-like avatar positively moderating the effect of positive displayed emotion on customer positive affect compared to displaying a computer-like avatar. The results indeed showed that a chatbot's display of positive emotions had a significantly positive effect on customer positive affect ($\beta = .216$, $p < .05$) when a human-like avatar is shown. However, this affective delivery did not affect customers' affective state when a computer-like avatar was shown. In this condition, no significant effect on customer positive affect was found ($\beta = -.003$, $p = .975$). To compare whether the differences between the groups were significant, once more the results from the permutation test were consulted. These results indicated no significant difference regarding the effect of positive emotions on customer positive affect between the human-like and the computer-like group ($p = .165$). Based on the permutation test, the decision had to be made to reject the proposed H6. Similar results could be obtained regarding H7, which proposed a moderating effect of a human-like avatar on the positive effect of positive displayed emotions on empathy. While in the human-like group displaying positive emotions had a positive and significant effect on empathy ($\beta = .300$, $p < .01$), this effect did not reach statistical significance in the computer-like group ($\beta = .164$, $p = .102$). However, as the permutation test indicated that the difference between the two groups was not significant ($p = .355$), H7 had to be rejected. Both models are presented in Figure 5.12 and Figure 5.13. The path coefficients, their significances, and the permutation results are summarized in Table 5.34. The table also contains the path coefficients and their significances for the investigated control variables, which were the same as in the previous studies (need for interaction, emotional decision behavior, experience with live chats, experience with chatbots, age, and gender).

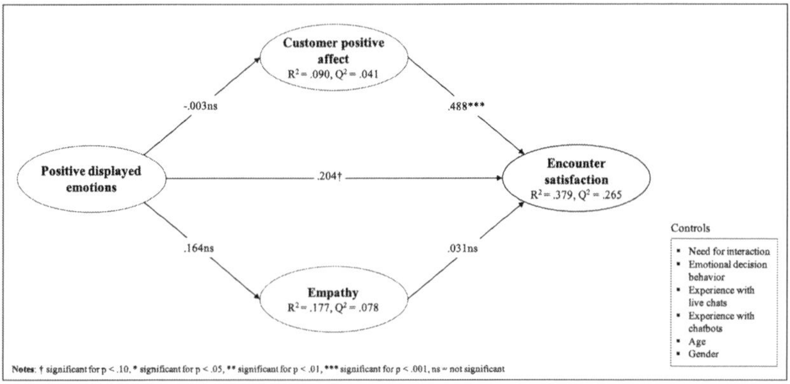

Figure 5.12 Results Structural Equation Modeling (Computer-Like), Study 5

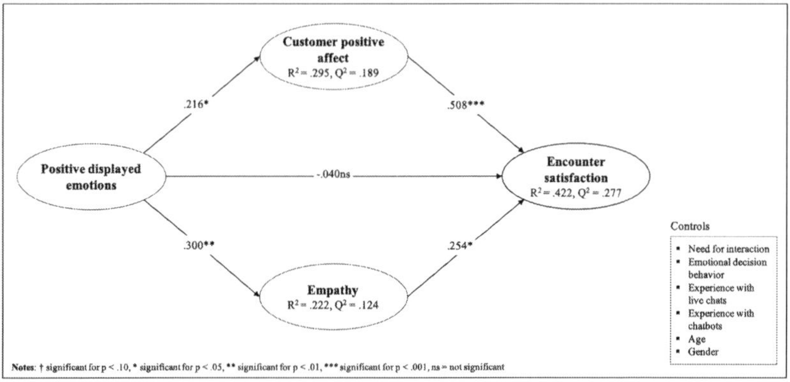

Figure 5.13 Results Structural Equation Modeling (Human-Like), Study 5

For various reasons, the results were nevertheless such that meaningful findings could be extracted regarding the effect of a human-like avatar. First, based on the results within the human-like group and within the computer-like group, a starting point could be obtained. While in one group (human-like) the effect of positive emotions on customer positive affect was positive and significant, it was not in the other group (computer-like). This fact strongly suggested that there must be avatar-dependent effects, as except for the displayed avatar, everything else was identical. The assumption was supported by the fact that in the human-like group, the variance of the construct customer positive affect could be explained to a greater extent by the representation of positive emotions than in the computer-like group. Second, the permutation approach is a conservative approach in terms of committing a Type I error (i.e., the null hypothesis is accepted, although there is a difference between the investigated groups) (Chin & Dibbern, 2010; Hair et al., 2018b). Third, it must be stated that considering the complexity of the evaluated research model, the sample must be classified as relatively small. In this context, the recommendation for sample size points toward 200 subjects (Kline, 2011). It must be mentioned that this could almost be met in the present context. However, the size was distributed over four experimental conditions. For this reason, it is likely to expect difficulties in extracting significant results in the present case. These different aspects motivated to investigate the impact of different avatars on the affective responses of customers in more detail. That there are effects resulting from the shown avatar becomes clear later in the course of the additional analyses, where this aspect is taken up again.

Looking at the effect sizes, the results presented before could be confirmed. While in the computer-like group the presentation of positive emotions had no effect on customer positive affect ($f^2 = .000$), a weak effect was found in the human like group ($f^2 = .058$). Likewise, the results shown regarding empathy could be confirmed. In the computer-like group, the effect of positive displayed emotions was substantially weaker ($f^2 = .029$) than in the human-like group ($f^2 = .097$). Table 5.35 shows the effect sizes of both subsamples.

Table 5.34 Standardized Path Coefficients and Significances, Study 5

	Computer-like avatar			Human-like avatar			Permutation
	Standardized coefficient	Standard deviation	T-statistics	Standardized coefficient	Standard deviation	T-statistics	p value
Main effects							
Positive displayed emotions ➜ customer positive affect	−.003	.112	.031	.216	.094	2.302*	.165
Positive displayed emotions ➜ empathy	.164	.100	1.637	.300	.096	3.139**	.355
Positive displayed emotions ➜ encounter satisfaction	.204	.107	1.914†	−.040	.088	.454	.074†
Customer positive affect ➜ encounter satisfaction	.488	.098	4.996***	.508	.120	4.219***	.907
Empathy ➜ encounter Satisfaction	.031	.135	.228	.254	.105	2.407*	.217
Indirect effects							

(continued)

Table 5.34 (continued)

	Computer-like avatar			Human-like avatar			Permutation
	Standardized coefficient	Standard deviation	T-statistics	Standardized coefficient	Standard deviation	T-statistics	p value
Positive displayed emotions ➜ customer positive affect ➜ encounter satisfaction	−.002	.056	.030	.110	.060	1.837†	.163
Positive displayed emotions ➜ empathy ➜ encounter satisfaction	.005	.027	.186	.076	.045	1.705†	.148
Control paths							
Need for interaction ➜ customer positive affect	−.041	.159	.255	−.247	.101	2.445*	.311
Need for interaction ➜ empathy	−.181	.144	1.262	−.285	.102	2.808**	.541

(continued)

Table 5.34 (continued)

	Computer-like avatar			Human-like avatar			Permutation
	Standardized coefficient	Standard deviation	T-statistics	Standardized coefficient	Standard deviation	T-statistics	p value
Need for interaction ➜ encounter satisfaction	−.144	.107	1.347	.124	.079	1.569	.032*
Emotional decision behavior ➜ customer positive affect	.046	.116	.398	.247	.097	2.531*	.176
Emotional decision behavior ➜ empathy	.087	.103	.841	.177	.106	1.665	.548
Emotional decision behavior ➜ encounter satisfaction	−.116	.101	1.156	.027	.089	.300	.298
Experience with live chats ➜ customer positive affect	−.330	.140	2.356*	.223	.107	2.089*	.001**

(continued)

Table 5.34 (continued)

	Computer-like avatar			Human-like avatar			Permutation
	Standardized coefficient	Standard deviation	T-statistics	Standardized coefficient	Standard deviation	T-statistics	p value
Experience with live chats → empathy	−.392	.140	2.800**	−.031	.137	.223	.059†
Experience with live chats → encounter satisfaction	−.042	.136	.311	.095	.120	.789	.457
Experience with chatbots → customer positive affect	.289	.146	1.980*	−.012	.114	.108	.127
Experience with chatbots → empathy	.402	.127	3.154**	.055	.145	.379	.087†
Experience with chatbots → encounter satisfaction	.012	.114	.108	.026	.115	.223	.940
Age → customer positive affect	−.074	.101	.729	.008	.092	.086	.544

(continued)

Table 5.34 (continued)

	Computer-like avatar			Human-like avatar			Permutation
	Standardized coefficient	Standard deviation	T-statistics	Standardized coefficient	Standard deviation	T-statistics	p value
Age ➜ empathy	−.034	.122	.276	.110	.093	1.175	.344
Age ➜ encounter satisfaction	.067	.105	.637	−.033	.083	.398	.447
Gender (female) ➜ customer positive affect	.023	.104	.217	.115	.105	1.089	.538
Gender (female) ➜ empathy	−.076	.103	.733	.019	.101	.186	.513
Gender (female) ➜ encounter satisfaction	.214	.082	2.611**	−.016	.095	.164	.082†

Notes: † significant for $p < .10$, * significant for $p < .05$, ** significant for $p < .01$, *** significant for $p < .001$.

5.6.6.7 Additional Analyses

To shed more light on the effect of a human-like avatar relative to a computer-like avatar, it is important to include a situation without an avatar. Therefore, the data of *Study 2* and *Study 4* was pooled with the two subsamples collected for *Study 5*. This was possible because the used stimuli during the experiment resembled each other except for the displayed avatars added to the stimuli for *Study 5*. The decision to test the two avatar conditions against the no avatar condition (taken from *Study 2* and *Study 4*) was primarily encouraged by the fact that the results shown above raised the question of whether a human-like avatar (i.e., the graphical human-like representation of a chatbot) enhances social responses by humans compared to a situation where only human-like behaviors are shown. At the same time, the question was raised whether the presentation of a computer-like avatar would attenuate the effects of the displayed positive emotions. To test the raised questions, two permutation-based MGAs were calculated. First, the human-like avatar group was tested against the control condition taken from *Study 2* and *Study 4* with no graphical representation. This procedure showed that the positive effects of positive displayed emotions on customer positive affect and on empathy were in the condition when no avatar was shown statistically not different from the effects in the human-like group ($p_{\text{customer positive affect}} = .732$, $p_{\text{empathy}} = .533$). Second, the same procedure was applied to test the computer-like group against the no avatar group. The results showed a significant difference between the two groups regarding the effect on customer positive affect ($p < .05$), however no significant difference concerning the effect on empathy ($p = .560$). These were important results, indicating that when positive emotions are displayed by a chatbot, no enhancing effects can be expected from a human-like avatar. By contrast, however, the effects of the affective delivery can be at least partially inhibited by the presentation of a computer-like avatar.

Table 5.35 Effect Sizes, Study 5

		2	3	4
Computer-like avatar				
Main effects				
1	Positive displayed emotions	.000	.029	.057
2	Customer positive affect			.268
3	Empathy			.001
4	Encounter satisfaction			

(continued)

Table 5.35 (continued)

		2	3	4
	Controls			
	Need for interaction	.002	.034	.027
	Emotional decision behavior	.002	.008	.018
	Experience with live chats	.063	.098	.001
	Experience with chatbots	.048	.102	.000
	Age	.005	.001	.006
	Gender (female)	.001	.006	.066
Human-like avatar				
Main effects				
1	Positive displayed emotions	.058	.097	.002
2	Customer positive affect			.286
3	Empathy			.079
4	Encounter satisfaction			
	Controls			
	Need for interaction	.085	.103	.023
	Emotional decision behavior	.079	.037	.001
	Experience with live chats	.042	.001	.009
	Experience with chatbots	.000	.003	.001
	Age	.000	.015	.002
	Gender (female)	.018	.000	.000

5.6.7 Discussion

Study 5 held several vital findings. The key insight was the uncovering of boundary conditions. The hypothesis that a more human-like graphical representation of the chatbot would positively influence the effects of displayed of positive emotions compared to a computer-like representation had to be rejected. The same results could be obtained when the human-like representation was tested against a situation where the chatbot did not have any graphical representation. This means that, contrary to the expectations, an additional human-like cue expressed through the human-like avatar did not enhance affective responses by the customers. This was different with respect to the computer-like avatar. In this respect, a clear negative effect could be found in the understanding that a computer-like representation of the chatbot prevented the occurrence of unconscious affective processes. These found effects naturally raise questions about possible causes.

Research on anthropomorphism could show that so-called cues, which promote the attribution of human features, are responsible for its occurrence (e.g., Blut et al., 2021; Hegel et al., 2008). Regarding these cues, however, it seems that the effect decreases with an increasing number of cues. This is indicated by the fact that in the study described above, no positive effect of a human-like avatar could be found compared to a situation without an avatar. This means, when a human-like cue in the form of positive displayed emotions was present, customers' affective responses could not be enhanced by the addition of a second human-like cue (expressed by the human-like avatar). Instead, however, these cues seem to have the potential to cancel each other out. In this context, the affective responses of the customers by expressing positive emotions were prevented by the presence of a cue that contradicted the behavior-based human-like perception of the chatbot. This is indicated by the fact that when a computer-like avatar was present, no emotional contagion happened.

Despite the unambiguous results up to this point regarding the affective capabilities of chatbots and their effects on service encounters, limitations of these results remain. These limitations concern the generalizability of the results. On a methodological level, the affective reactions of customers to the expression of positive emotions by a chatbot could be demonstrated in different ways. On a content level, however, the procedure was associated with the limitation that a booking scenario always formed the starting point. This issue was additionally raised by the fact that the stimuli used were almost entirely identical in *Study 2*, *Study 4*, and *Study 5*. The only changing aspect was the used avatar for *Study 5*. In order to specifically address this limitation and to shed light on whether the affective processes are also applicable in other situations, *Study 6*, presented below, was developed. In addition to the change of the service context, *Study 6* focused on the examination of the effects caused by positive displayed emotions depending on a successful or an unsuccessful service recovery.

5.7 Study 6: Emotional Contagion in Service Recovery Encounters[6]

5.7.1 Conceptual Development

Up to this point, the results regarding emotional contagion from chatbot to a human in the context of a service encounter were unambiguous and in line with derived expectations from human-to-human interactions. This means that a chatbot displaying positive emotions can cause emotional contagion, resulting in a better service evaluation. Expectations were also confirmed concerning empathy toward the chatbot's affective state. The same display of positive emotions does also cause stronger anthropomorphic reactions toward that chatbot, resulting in customers better being able to empathize with the chatbot. The results provide clear support that this reaction enhances service evaluations. In particular, the positive effects on the evaluation of the service should be noted. This is important because one of the primary goals is to provide good service, especially in the nowadays highly competition-driven markets (Lee, Singh, & Chan, 2011). However, in the delivery of services, there may inevitably occur failures, meaning situations where there is a mismatch between customer expectations and the service delivered (La & Choi, 2019). Service failures regularly require firms to take action to correct those mismatches. This is when service recoveries come into play. Johnston and Hewa (1997, p. 467) define service recoveries as "[...] the actions of a service provider to mitigate and/or repair the damage to a customer that results from the provider's failure to deliver a service as it is designed." As AI continues to spread in service encounters, it becomes inevitable that AI will play an active part in recovery actions. For this reason, De Cicco et al. (2020) call for the use of AI in recovery situations to become more prominent in scientific research.

Generally, the relevance of considering the influence of positive displayed emotions in these situations arises from the fact that there is still the possibility of satisfying the customer, although the service failure has occurred (McCollough & Sundar, 1992). Research has shown that the occurrence of a service failure usually triggers negative emotions (Patterson, McColl-Kennedy, Smith, & Lu, 2009; Smith & Bolton, 2002; Zhang, Zhang, & Sakulsinlapakorn, 2020). Research on impact factors of customer satisfaction in recovery encounters has found that one influencing factor can be the aspect of time (del Río-Lanza, Vázquez-Casielles, & Díaz-Martín, 2009; Maxham III & Netemeyer, 2002). On this basis, it can be

[6] Parts of *Study 6* have been presented at the 7th Rostock Conference on Service Research.

assumed that if a customer turns to a company after a service failure and the company is unable to rectify the failure, the negative affect may additionally be intensified because the customer needs to spend additional time and effort to have the failure rectified. In contrast, if a company is able to resolve the inconveniences caused at the first attempt, this possibly has the effect of reducing the negative emotions present on the part of the customer. From this, a positive effect of a successful recovery situation is expected on the central variables of the proposed research model when it is compared to a situation when recovery ends unsuccessfully (Figure 5.14).

H8a: A successful course of a recovery situation has a positive effect on customer positive affect compared to an unsuccessful recovery.
H8b: A successful course of a recovery situation has a positive effect on empathy compared to an unsuccessful recovery.
H8c: A successful course of a recovery situation has a positive effect on recovery satisfaction compared to an unsuccessful recovery.

In addition to reducing the customer's negative emotional state by providing an immediate solution to the problem that has arisen, the transfer of positive emotions can also contribute to mitigate customers' negative affect. Motivation for the assumption of emotional contagion and the favorable effects on the evaluation of the specific service interaction comes from Du et al. (2011). In their experiment, they were able to show that the expression of positive emotions by an employee can reduce negative emotions on the part of customers. In another context, Smith and Bolton (2002) pointed out that customers' emotional reactions may well play a central role in service recoveries in terms of the evaluation of a company's efforts to rectify the failure. They were able to show that present negative emotions are decisive for the justice evaluation, playing a critical role in recovery situations.

In the course of research around service recoveries, the focus has increasingly shifted to perceived justice in recent years. Research could show that it is heavily dependent on the perception that one has been treated fairly, which leads to satisfaction with a recovery situation (La & Choi, 2019). Neuroscience currently assumes that the brain's reward system is behind the feeling of justice (Buckholtz & Marois, 2012; Ruff, 2014). This reward system is triggered by a social cognition system, which in social interactions evaluates, among other things, the extent to which the other person's cooperative intentions are perceived. Various authors view justice as a cognitively shaped evaluation of the situation at hand (e.g., Smith & Bolton, 2002; Smith, Bolton, & Wagner, 1999). For other authors (e.g., Baron, Harris, Elliott, Schoefer, & Ennew, 2005), justice perception plays the role of a cognitive counterpart to the affective state in evaluating recovery satisfaction. Thus, it becomes clear, there is high cognitive effort behind the justice assessment. The conjecture is supported by the fact that Smith and Bolton (2002) were able to show that the negative affect felt by customers in the aftermath of a failure leads to the evaluation being comparatively more systematic at the cognitive level and less influenced by heuristics (e.g., affect-as-information).

In general, the same affective responses by customers to a chatbot's affective delivery that were investigated throughout the previous studies are expected to occur also in a service recovery situation. What is different, is the outcome variable. Where previously encounter satisfaction was investigated, now recovery satisfaction is of interest. In sum, this means, customer positive affect is expected to mediate the relationship between positive displayed emotions by a chatbot and recovery satisfaction. The same holds for empathy. Regarding the latter, however, effects emerging from the course of the recovery are assumed because both recovery situations (unsuccessful and successful) lead to the need for a great deal of cognitive effort, which is expected to be higher in unsuccessful recovery situations. This is because in the course of a successful service recovery, the influencing factors associated with justice, such as the compensation, are usually comparatively easy to access and thus are expected to require less cognitive effort. If the solution to the problem that has arisen cannot be offered directly but the customer is referred to another instance, much more effort is required to determine the extent to which the behavior and settlement correspond to the justice and fairness expectations. Ultimately, this means that the cognitive resources necessary to expend in the course of an unsuccessful recovery lead to a diminished ability to empathize with the emotional perspective of the chatbot. Due to the cognitive resources required for the evaluation of the recovery encounter, it is expected that it is more difficult for customers to empathize with the chatbot if the recovery ends unsuccessfully. Therefore, it is expected that a successful course of a recovery encounter positively moderates the positive effect of positive displayed emotions on empathy. This means, as less cognitive effort is required in successful recovery encounters to evaluate the treatment and for justice perceptions, more cognitive resources are left to empathize with the chatbot. Accordingly, the following is proposed (Figure 5.14):

H9: A successful recovery situation positively moderates the effect of positive displayed emotions on empathy.

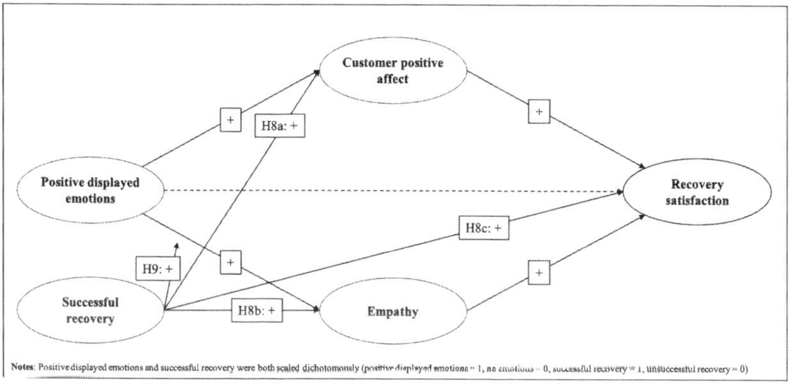

Figure 5.14 Proposed Research Model and Hypotheses, Study 6

5.7.2 Research Objectives and Setting

The objective of the present study was to address service recovery situations and to examine if and to what extent the effects around emotional contagion and empathy are affected by the success of recovery efforts. The change in the service situation meant new service scenarios had to be developed. For reasons of consistency, an attempt was made to design a recovery situation in a hotel context. In the conception of the scenario, it should be ensured that, on the one hand, the failure did not take place in the situation under consideration itself so the situation could focus solely on the recovery attempts. On the other hand, the failure itself should not be associated with the chatbot to exclude that the chatbot itself is already afflicted with negative experiences. For both aspects, the intervening effects needed to be reduced to a minimum to address the question of emotional contagion, empathy, and the expression of positive emotions in service recoveries. An erroneous room assignment of a hotel guest appeared to be suitable for the conception, which made it possible to use a hotel setting again and with this to rule out possible disturbances due to the setting. To account for a possible feeling of strangeness due to having to conduct such a service request via a chat, the study's opening credits made special reference to the fact that the fictive hotel was one with a strong focus on the use of digital technologies.

5.7.3 Experimental Design

The manipulation was a 2×2 between-subjects design. The manipulated factors were positive displayed emotions on the one hand and the success of the service recovery on the other hand. In line with previous studies, to manipulate positive emotions, the 😀-emoji was again combined with witty statements by the chatbot. Success of the service recovery was manipulated by the fact that in the successful scenarios, the chatbot was able to provide a suggestion on how to solve the problem to the customer and thus solve the problem directly. In the unsuccessful scenarios, the chatbot had to admit that it could not directly help with the problem, so the chatbot referred the customer to the customer service being available via telephone. Participants randomly were assigned to one of the four scenarios.

The starting point for the recovery was the service failure that the customer had booked a double room and, upon entering his room, found that he had been given a single room. The design of the stimuli was based as much as possible on the previous stimuli. Therefore, all videos started with the customer being welcomed by the chatbot shortly after entering the fictitious chat. As before, existing implications from chatbot research regarding the design were considered, so the chatbot again presented its capabilities together with the greeting. The different progression of scenarios began with the customer asking if the chatbot could assist with a complaint. While the chatbot affirmed this question in the successful scenarios and provided the customer with the suggestion of a rebooking, the chatbot negated the question in the unsuccessful scenarios and referred the customer to the customer service available by telephone. In the hotel industry, it is common practice to offer customers a free room upgrade when service failures occur (Lee et al., 2011). In designing the stimuli, this aspect served to reinforce the manipulation of the successful recovery by having the chatbot offering the customer a double room rated one category higher without being charged any extra money. Both conditions ended largely identically, with the customer thanking the chatbot for the information (and, in the successful condition, for the settlement). The chat ended with a final message from the chatbot, which then went offline. This followed the procedure of the previous studies. The duration of the fictive encounters was approximately the same length as the booking encounters of the previous studies.

For the manipulation of positive emotions, the results from the previously conducted studies served as an orientation point. The placement of the 😀-emoji followed the previous studies where possible. For the formulation of confirmations or other statements of the chatbot, the formulations from previous studies were used, unless a change was necessary. Wit was brought into play insofar as the chatbot made a statement toward the end of the chats that it could be helpful in the search for excursion destinations. In the manipulated scenarios, this statement was supplemented by the hint that the locals would know the city's secret corners after all.

Essential elements of the videos remained unchanged. This applied to aspects like the answering time. The videos again had a size of 1920 pixels (height) by 1080 pixels (width) and were created using Adobe After Effects v16.1.1 for MacOS.

5.7.4 Procedure

Based on the positive and reliable experience with the chosen procedure for the data collection during the previous studies, the subjects for the present study were also acquired via Clickworker. Again, they received monetary compensation for their participation. The structure of the study was identical to that of the first four web-based studies. Satisfaction with the service recovery was measured immediately after empathy toward the chatbot.

5.7.5 Measures

The scales used for the study to measure the constructs were the same as the measurement instruments used for the previous studies. Only for evaluating satisfaction with the recovery efforts, an additional scale was adapted, which is presented below. In the research model, recovery satisfaction replaced encounter satisfaction. Table 5.36 provides an overview of all constructs.

Recovery satisfaction

To measure satisfaction with the undertaken service recovery efforts, a scale by La and Choi (2019) was adapted, which consisted of two items. Compared to general satisfaction with the service experienced, this scale was more strongly focused on the efforts to remedy the previous failure. The items used were "Overall, I was satisfied with how my complaint was handled" and "In general, I was satisfied with the outcome that I received." As with all other items, the two items were measured using a seven-point Likert-type scale ranging from "completely disagree" (1) to "agree completely" (7) (CA = .969, CR = .985, AVE = .970).

Table 5.36 Items and Indicator Loadings for Latent Constructs, Study 6

Construct	Item loading	CA	CR	AVE
Main constructs				
Customer positive affect		.970	.975	.868
I feel strong.	.894			
I feel elated.	.946			
I feel excited.	.947			
I feel enthusiastic.	.928			
I feel active.	.929			
I feel peppy.	.946			
Empathy		.875	.922	.798
I was very close to the chatbot's emotions.	.922			
I felt compassionate toward the chatbot.	.893			
I felt soft-hearted toward the chatbot.	.862			
Recovery satisfaction		.969	.985	.970
Overall, I was satisfied with how my complaint was handled.	.985			
In general, I was satisfied with the outcome that I received.	.985			
Controls				
Need for interaction		.932	.957	.881
I feel more relaxed when I have personal contact with service employees.	.909			
I rather like to communicate with people when services are provided.	.956			
I prefer personal contact for asking and answering my questions.	.949			
Emotional decision behavior		.896	.951	.906
In general, in decision making, I tend to be guided by my rational side/my emotional side.	.950			
In general, in decision making, I tend to be guided by my head/my heart.	.953			

Notes: CA = Cronbach's alpha, CR = composite reliability, AVE = average variance extracted

5.7.6 Participants

In the original sample, 322 participants were contained. However, subjects were removed if their data were incomplete ($n = 5$) or if they had answered the check for attention wrong ($n = 38$). Further 58 subjects were removed because they seemed to not have recognized the experimental treatment. The final sample of the conducted investigation consisted of 221 participants. From this, 57.5 percent ($n = 127$) were male and 42.1 percent were female ($n = 93$). One participant reported being neither male nor female. Participants were on average 38.35 years old ($SD = 12.30$ years).

5.7.7 Results

5.7.7.1 Common-method Bias

To address the problem of CMV, following the previous studies, the same a priori and statistical remedies were applied. These included as a priori remedies, for example, measuring independent and dependent constructs on different pages. As statistical remedy, the CLC approach was applied (Chin et al., 2013). For this, the blue attitude scale (Simmering et al., 2015) already introduced in *Study 3* was used.

5.7.7.2 Treatment Check

Following the first studies' procedure, participants were removed from the sample if their answers indicated they had not recognized the experimental treatment (Sigall & Mills, 1998). Subjects were excluded if they rated the displayed emotions as being extraordinarily positive while being assigned to one of the control conditions (i.e., without positive emotions). Similarly, they were excluded if they perceived the expressed emotions as being very low in their positivity while being assigned to a condition with positive emotions being shown. Before removing any participants, a one-way between-subjects ANOVA was calculated to check if the manipulated emotions in principle were recognized as being positive. This procedure revealed a significant effect across the four involved scenarios ($F(3, 275) = 18.761$, $p < .001$). According to the Levene's test, homogeneity of variances was established ($p = .860$). In the further course, LSD post hoc tests were calculated to compare the groups. The test showed for the participants of the successful manipulation group that those who were exposed to positive emotions scored significantly higher in terms of the perceived positivity ($M_{manipulation} = 4.888$, $SD = 1.360$) than those who were exposed to the control situation ($M_{control} =$

4.259, SD = 1.481, $p < .01$). Those participants who were assigned to one of the unsuccessful scenarios reported significantly more positivity when exposed to positive emotions ($M_{manipulation}$ = 3.631, SD = 1.447) than did those who saw the chatbot communicating with no positive emotions ($M_{control}$ = 3.214, SD = 1.384, $p < .10$).

5.7.7.3 Path Model Estimation

SmartPLS v3.3.3 was used again to calculate the proposed model. Both experimental variables were added to the model as dichotomously scaled variable. While the expression of positive emotions followed the previous studies (positive displayed emotions = 1, no emotions = 0), successful recovery was scaled as follows: 1 = successful recovery, 0 = unsuccessful recovery. To test the proposed moderation of a successful recovery on the effect of positive displayed emotions on empathy, the interaction term of positive displayed emotions and successful recovery was calculated and added to the model. Inference tests were calculated using a bootstrapping procedure with 5,000 resamplings.

5.7.7.4 Measurement Model

The results of a factor analysis, extracted from SmartPLS, were consulted to evaluate the reliability of the used indicators. This showed that all indicators loaded at least .862 on their respective factor and were thus above the critical threshold of .70 (Hulland, 1999) (Table 5.36). Hence, indicator reliability could be considered as being approved. As before, internal consistency was evaluated using CR. CR was equal to or above .922 for all constructs and therefore exceeded the suggested threshold of .70 (Bagozzi & Yi, 1988) (Table 5.36). Accordingly, internally consistent results could be assumed. This also held true for convergent validity. The AVE was at least equal to or above .798 (Table 5.36). The AVEs thus were above the threshold of .50 (Bagozzi & Yi, 1988). Discriminant validity was again assessed, consulting the cross loadings. No violation could be found as the indicators loaded highest on their intended construct (Table 5.37). The Fornell-Larcker criterion confirmed no violation against discriminant validity being present, as the square root of the AVEs exceeded the correlations with every other construct in all cases (Table 5.38). Lastly, the HTMT method showed a maximum ratio of .790, indicating no violation against discriminant validity even when compared to the conservative threshold of .85 (Kline, 2011) (Table 5.39).

Table 5.37 Discriminant Validity (Cross Loadings), Study 6

	PDE	SR	CPA	EP	RS	NFI	EDB	EXPLC	EXPCB	Age	Gender
PDE	*1.000*	.004	.173	.296	.110	−.214	.112	.120	.077	−.207	−.061
SR	.004	*1.000*	.681	.524	.758	−.176	.014	−.012	.019	−.117	.107
CPA_01	.105	.617	*.894*	.614	.769	−.231	.015	.063	.096	−.208	.027
CPA_02	.184	.660	*.946*	.658	.745	−.306	.079	.123	.170	−.229	.055
CPA_03	.164	.694	*.947*	.689	.786	−.321	.108	.109	.112	−.233	.079
CPA_04	.202	.610	*.928*	.712	.682	−.284	.125	.156	.150	−.297	.112
CPA_05	.137	.577	*.929*	.619	.632	−.312	.035	.125	.131	−.126	.062
CPA_06	.172	.634	*.946*	.662	.669	−.336	.100	.125	.148	−.191	.089
EP_01	.274	.440	.617	*.922*	.502	−.256	.215	.043	.054	−.189	.077
EP_02	.311	.381	.560	*.893*	.449	−.273	.190	.025	.025	−.181	.109
EP_03	.221	.549	.691	*.862*	.733	−.282	.088	.145	.178	−.215	.086
RS_01	.121	.741	.745	.636	*.985*	−.223	−.002	−.031	.046	−.138	.036
RS_02	.096	.751	.770	.645	*.985*	−.219	.031	−.030	.036	−.159	.081
NFI_01	−.176	−.116	−.293	−.239	−.186	*.909*	−.035	−.013	−.035	.063	.039
NFI_02	−.188	−.202	−.296	−.296	−.224	*.956*	.004	−.078	−.056	.152	.031
NFI_03	−.233	−.173	−.312	−.317	−.219	*.949*	.011	−.068	−.054	.122	.022
EDB_01	.109	.014	.087	.158	.025	.006	*.950*	.009	.071	−.085	.205

(continued)

Table 5.37 (continued)

	PDE	SR	CPA	EP	RS	NFI	EDB	EXPLC	EXPCB	Age	Gender
EDB_02	.104	.013	.073	.178	.003	−.017	*.953*	−.003	.044	−.101	.196
EXPLC	.120	−.012	.125	.088	−.031	−.058	.003	*1.000*	.656	−.188	.083
EXPCB	.077	.019	.144	.107	.041	−.052	.060	.656	*1.000*	−.093	.072
Age	−.207	−.117	−.231	−.221	−.151	.122	−.098	−.188	−.093	*1.000*	−.021
Gender	−.061	.107	.076	.101	.059	.032	.211	.083	.072	−.021	*1.000*

Notes: PDE = positive displayed emotions, SR = successful recovery, CPA = customer positive affect, EP = empathy, RS = recovery satisfaction, NFI = need for interaction, EDB = emotional decision behavior, EXPLC = experience with live chats, EXPCB = experience with chatbots

Table 5.38 Discriminant Validity (Fornell-Larcker Criterion), Study 6

	Latent construct	1	2	3	4	5
1	Customer positive affect					
2	Empathy	.708				
3	Recovery satisfaction	.769	.650			
4	Need for interaction	−320	−.305	−.224		
5	Emotional decision behavior	.084	.177	.015	−.006	
	Square root of the AVE	.932	.893	.985	.938	.952

Notes: Scores indicate the correlation between the constructs

Table 5.39 Discriminant Validity (HTMT), Study 6

		1	2	3	4	5	6	7	8	9	10	11
1	Positive displayed emotions											
2	Successful recovery	.004										
3	Customer positive affect	.175	.689									
4	Empathy	.321	.546	.755								
5	Recovery satisfaction	.112	.770	.790	.681							
6	Need for interaction	.220	.181	.337	.332	.235						
7	Emotional decision behavior	.118	.015	.089	.207	.023	.026					
8	Experience with live chats	.120	.012	.127	.085	.031	.058	.006				
9	Experience with chatbots	.077	.019	.147	.103	.042	.053	.064	.656			
10	Age	.207	.117	.233	.233	.153	.124	.103	.188	.093		
11	Gender	.061	.107	.077	.108	.060	.034	.223	.083	.072	.021	

5.7.7.5 Structural Model

Following the investigation of the measurement model, the structural model was checked according to the recommendations by Hair et al. (2012). It can be determined that the model was able to account for a good portion of the variance of

Table 5.40 Descriptives and Correlations, Study 6

		M/%	SD	1	2	3	4	5	6	7	8	9	10	11
1	Positive displayed emotions	51.1%	–											
2	Successful recovery	52.0%	–	.004										
3	Customer positive affect	4.411	1.680	.173	.681									
4	Empathy	3.833	1.553	.296	.524	.708								
5	Recovery satisfaction	5.218	1.935	.110	.758	.769	.650							
6	Need for interaction	5.590	1.389	−.214	−.176	−.320	−.305	−.224						
7	Emotional decision behavior	3.310	1.493	.112	.014	.084	.177	.015	−.006					
8	Experience with live chats	5.548	1.521	.120	−.012	.125	.088	−.031	−.058	.003				

(continued)

Table 5.40 (continued)

		M/%	SD	1	2	3	4	5	6	7	8	9	10	11
9	Experience with chatbots	4.955	1.543	.077	.019	.144	.107	.041	−.052	.060	.656			
10	Age	38.350	11.300	−.207	−.117	−.231	−.221	−.151	.122	−.098	−.188	−.093		
11	Gender (female)	42.1%	–	−.062	.107	.076	.101	.060	.032	.211	.083	.072	−.021	

Notes: Correlations equal to or above |.144| are statistically significant ($p < .05$, two-tailed).

the investigated constructs. The average R^2 was .564. The Q^2 values in all cases exceeded 0 (Chin, 1998; Henseler et al., 2009), which indicated that the model possessed predictive relevance (Figure 5.15). As the VIFs did not exceed 2.919, collinearity was, according to Hair et al. (2019) who suggest 3 as a threshold, considered as being of no concern (Table 5.41).

Table 5.41 Variance Inflation Factors, Study 6

		3	4	5
Main effects				
1	Positive displayed emotions	1.120	1.129	1.211
2	Successful recovery	1.068	1.071	2.000
3	Customer positive affect			2.919
4	Empathy			2.266
5	Recovery satisfaction			
	Moderating effect			
	Positive displayed emotions x successful recovery		1.042	
Controls				
	Need for interaction	1.091	1.090	1.163
	Emotional decision behavior	1.084	1.116	1.110
	Experience with live chats	1.875	1.929	1.884
	Experience with chatbots	1.807	1.834	1.818
	Age	1.125	1.163	1.148
	Gender	1.079	1.087	1.082

5.7.7.6 Hypothesis Testing

H8a assumed a positive effect of a successful service recovery on customer positive affect. This hypothesis could be confirmed ($\beta = .636$, $p < .001$). The same applied to the positive effect of a successful service recovery on empathy ($\beta = .480$, $p < .001$) and recovery satisfaction ($\beta = .400$, $p < .001$), which delivered support for H8b and H8c. The chosen procedure to add successful recovery as a variable to the model opened the possibility of evaluating to what extent the effects around emotional contagion and empathy also apply in the context of service recoveries. This showed, in line with the previously presented studies, displaying positive emotions positively affected customers' positive affect ($\beta = .096$, $p < .05$), which in turn influenced satisfaction with the recovery ($\beta = .378$, p < .001). Furthermore, the effects concerning empathy could be confirmed. Displaying positive emotions significantly impacted empathy (β = .229, p < .001), which in the further course positively influenced recovery satisfaction (β = .196, p < .01) Both indirect effects were statistically significant while the direct effect lost its statistical significance when the mediators were inserted (β = .005, p = .893). This means that customer positive affect (positive displayed emotions ➜ customer positive affect ➜ encounter satisfaction = .036, p < .10) and empathy (positive displayed emotions ➜ empathy ➜ encounter satisfaction = .045, p < .05) fully mediated the relationship between positive displayed emotions and recovery satisfaction (Figure 5.15). The unconscious indirect effect through customer positive affect accounted for 45% of the total effect, whereas the conscious process through empathy accounted for the remaining 55%.

In a second step, the moderating effect of a successfully performed service recovery was investigated. In contrast to what was hypothesized in H9, the successful handling of a service failure during a recovery situation did not positively moderate the effect of positive displayed emotions on empathy (β = .066, p = .225). Therefore, H9 had to be rejected. All path coefficients, including the insignificant ones, are reported in Table 5.42.

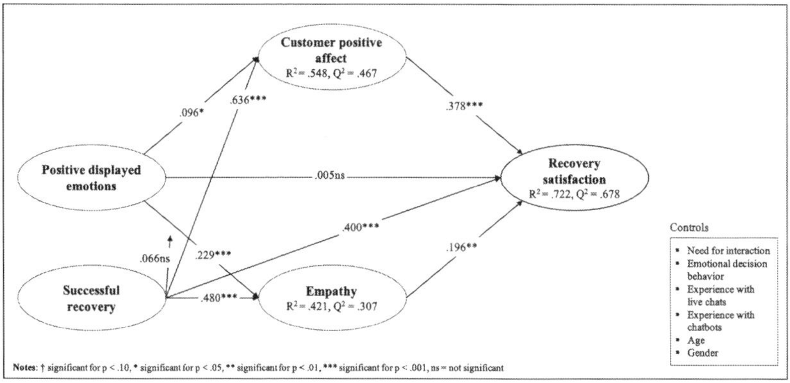

Figure 5.15 Results Structural Equation Modeling, Study 6

Consulting the f^2 effect sizes, the effects shown above could be confirmed. A successful recovery had a strong effect on customer positive affect (f^2 = .839) and empathy (f^2 = .371) while it had a moderate effect on recovery satisfaction (f^2 = .289). Opposing results could be obtained regarding the effects caused by the display of positive emotions, which had no effect on customer positive affect ($f^2 = .018$) and a weak effect on empathy ($f^2 = .080$). Furthermore, regarding the impacts on recovery satisfaction, customer positive affect had a moderate effect ($f^2 = .176$) while empathy had a weak effect ($f^2 = .061$). Table 5.43 provides an overview of all effect sizes.

Table 5.42 Standardized Path Coefficients and Significances, Study 6

	Standardized coefficient	Standard deviation	T-statistics
Main effects			
Positive displayed emotions ➜ customer positive affect	.096	.049	1.965*
Positive displayed emotions ➜ empathy	.229	.056	4.099***
Positive displayed emotions ➜ recovery satisfaction	.005	.039	.134
Customer positive affect ➜ recovery satisfaction	.378	.081	4.655***
Empathy ➜ recovery satisfaction	.196	.060	3.238**
Successful recovery ➜ customer positive affect	.636	.042	15.097***
Successful recovery ➜ empathy	.480	.049	9.789***
Successful recovery ➜ recovery satisfaction	.400	.060	6.713***
Indirect effects			
Positive displayed emotions ➜ customer positive affect ➜ recovery satisfaction	.036	.021	1.716†
Positive displayed emotions ➜ empathy ➜ recovery satisfaction	.045	.017	2.559*
Moderating effect			
Positive displayed emotions x successful recovery ➜ empathy	.066	.054	1.215

(continued)

Table 5.42 (continued)

	Standardized coefficient	Standard deviation	T-statistics
Control paths			
Need for interaction ➜ customer positive affect	−.168	.043	3.943***
Need for interaction ➜ empathy	−.159	.056	2.834**
Need for interaction ➜ recovery satisfaction	.025	.038	.642
Emotional decision behavior ➜ customer positive affect	.051	.050	1.020
Emotional decision behavior ➜ empathy	.118	.059	1.992*
Emotional decision behavior ➜ recovery satisfaction	−.055	.041	1.332
Experience with live chats ➜ customer positive affect	.048	.057	.837
Experience with live chats ➜ empathy	.003	.080	.033
Experience with live chats ➜ recovery satisfaction	−.107	.048	2.249*
Experience with chatbots ➜ customer positive affect	.072	.059	1.205
Experience with chatbots ➜ empathy	.050	.074	.668
Experience with chatbots ➜ recovery satisfaction	.033	.049	.669
Age ➜ customer positive affect	−.093	.051	1.811†
Age ➜ empathy	−.074	.064	1.143
Age ➜ recovery satisfaction	.006	.046	.136
Gender (female) ➜ customer positive affect	−.003	.048	.057
Gender (female) ➜ empathy	.031	.053	.575
Gender (female) ➜ recovery satisfaction	−.014	.038	.363

Notes: † significant for $p < .10$, * significant for $p < .05$, ** significant for $p < .01$, *** significant for $p < .001$.

Table 5.43 Effect Sizes, Study 6

		3	4	5
Main effects				
1	Positive displayed emotions	.018	.080	.000
2	Successful recovery	.839	.371	.289
3	Customer positive affect			.176
4	Empathy			.061
5	Recovery satisfaction			
Moderating effect				
	Positive displayed emotions x successful recovery		.007	
Controls				
	Need for interaction	.057	.040	.002
	Emotional decision behavior	.005	.022	.010
	Experience with live chats	.003	.000	.022
	Experience with chatbots	.006	.002	.002
	Age	.017	.008	.000
	Gender (female)	.000	.002	.001

5.7.8 Discussion

The results of the sixth study hold vital findings. First, the change from a booking scenario (i.e., a routine service encounter) to a recovery situation, which was preceded by a service failure, showed that the effects of a chatbot's affective delivery are not limited to routine service encounters. Thus, the study helps to increase the generalizability of the previous results and the discussed key findings. The unconscious process of emotional contagion remained present through the course of a service recovery. This is mainly important as it contains support for the validity of the affect-as-information theory, which in the present case means that a change in customers' positive affect during a service recovery did also lead to a better rating of the satisfaction with the undertaken recovery efforts. This fact greatly expands the significance of the presented results not only with regard to the occurrence of affect-related processes but also with regard to their relevance for service research. In this context, however, it is important to note that the results suggest that the effects studied are weaker. This seems to be true at least for the unconscious path through customer positive affect (i.e., emotional contagion). Although it seems that the expression of positive emotions by chatbots in service recoveries can make a positive contribution, the question arises as to whether customers expect other emotional reactions from a chatbot in such service encounters, for example, empathy or courtesy (Hocutt, Bowers, & Donavan, 2006).

Second, the results could show that the expected moderating effects caused by a successful or unsuccessful course of a recovery situation on the effect of positive displayed emotions on empathy could not be confirmed. From this, it can be concluded that the expected cognitive effort for the evaluation of the recovery situation does not significantly affect the ability to empathize with the chatbot. At the same time, this indicates that customers attribute human characteristics to chatbots even in critical service encounters. This process is not influenced by the course of the encounter. Although no final explanation could be provided why empathy toward the chatbot maintained its strong influence, the following aspects can serve as an explanatory approach. In service transactions, empathizing with the affective state of a counterpart also serves to inform the customer about the actual affective state and the abilities of an employee (Gremler & Gwinner, 2000). The previous results indicate that this fact can be similarly applied to chatbots. This means, as customers perceive chatbots as social actors and apply social rules, they do also assess the appropriateness of the displayed emotions and the abilities of the chatbots when interacting. Thus, while it is already important to carefully design a chatbot's emotional behavior in regular service encounters so that the emotions perceived by the customer are evaluated as authentic and sincere, this appears to be even more important in recovery situations, especially when help cannot be offered immediately. Therefore, the study is of particular relevance for managers. This will be discussed in detail later.

General Discussion

6

The question raised at the beginning of this thesis was how the increasing use of chatbots to handle service situations affects the emotional components of a service encounter. The phenomenon identified as the central object of the evaluation was emotional contagion (Hatfield et al., 1992), which in combination with the affect-as-information theory (Clore et al., 2001), could show that the affective delivery of an employee can lead to a change in the affective state of a customer. This ultimately results in a more positive service evaluation (e.g., Pugh, 2001). Furthermore, empathy was of interest, which was considered a cognitive counterpart to an emotional contagion (e.g., Decety & Lamm, 2006; Prochazkova & Kret, 2017). Based on the findings from human-to-human research, research on artificial entities, and computer-mediated communication, several research gaps were identified. In a series of experiments, both in the laboratory and web-based, those research gaps were closed, and two additional experiments explored boundary conditions.

The first research gap concerned emotional contagion and the question of whether a chatbot that can only rely on text-based communication can trigger emotional contagion. The presented results indicate that a transfer of emotions from a chatbot to a human is possible, following processes already known from human-based research via mimicry and feedback (Barsade, 2002). In particular, the laboratory study conducted at the beginning, in which customers' facial reactions were recorded and analyzed, provided clear support for this. The following studies were able to confirm these results several times. This yields great potential in accordance with the CASA paradigm (Reeves & Nass, 1996), but also the necessity to consider these aspects in the interaction with chatbots (e.g., when chatbots are developed).

The second research gap concerned chatbot-caused emotional contagion in a service context, as it is well documented from human-to-human research that

K. Prinz, *The Smiling Chatbot*,
https://doi.org/10.1007/978-3-658-40028-6_6

emotional contagion possesses high relevance for service research (e.g., Pugh, 2001). The results show that chatbot-handled service encounters follow similar rules. This is because, with regard to the change in affective state on the customer side, the validity of the affect-as-information theory (Clore et al., 2001) could be confirmed in chatbot-handled service encounters. This clearly shows that, contrary to many opinions that highlight the advantages on a technical level (e.g., Behera, 2016; Li et al., 2017; Schanke et al., 2021; Schneider, 2017; Wünderlich & Paluch, 2017), satisfaction evaluations in chatbot-handled service encounters are not purely functionally driven. Instead, emotions play a central role there as well, and humans apply social rules known from human-to-human interactions.

Third, it was identified that a critical research gap existed regarding anthropomorphic reactions of humans toward chatbots as disembodied conversational agents. In this regard, the chatbot's behavior (i.e., the display of positive emotions), which followed typical human interaction patterns, also led people to ascribe more human attributes to the chatbot in the form of its own affective state. They were also able to more strongly empathize with the chatbot. Showing human behavioral patterns by the chatbots in this context serves as a human-like cue. However, it should be emphasized that existing research in this regard has always strongly assumed physical characteristics or graphic representations to be predominantly responsible for such an attribution (e.g., von der Pütten et al., 2010). This thesis counters that the behavioral properties of a chatbot can trigger the same effects even without any physical or graphical representation at all. In this context, a notable point is that customers' affective responses depend, at least in part, on the personality traits of the customers. Extraverted customers were found to be more susceptible to catching the chatbot's positive emotions. The expectation that customers who are characterized by a high degree of openness to experience would be better able to empathize with the chatbot could, however, not be confirmed. Nevertheless, it turns out that not all customers react the same way to the affective delivery of the chatbot.

After investigating the basic mechanisms around emotional contagion and empathy, the fourth research gap pertained to how graphical and behavior-based human-like cues behave in combination. This formed the starting point for investigating boundary conditions for the affective processes previously examined with two further studies. Essential results could be found regarding the use of avatars to graphically represent chatbots. The tendency of humans to anthropomorphize artificial objects, such as chatbots, is usually enhanced by the representation of human-like cues such as a human-like appearance (Blut et al., 2021; Hegel et al., 2008). However, in the case of the considered chatbot, no positive moderating effect of a human-like avatar could be found. At the same time, however, the presentation of a computer-like avatar prevented the emergence of the investigated unconscious affective responses to the chatbot's behavior. On the one hand, the results strongly indicate a decreasing marginal benefit of adding additional human-like cues. On the other hand, the results also suggest that the simultaneous presentation of computer-like cues can inhibit the positive effects of behavior-based human-like cues.

Finally, customers' exhibited reservations toward chatbots reported in extant research (e.g., Dietvorst et al., 2015; Mozafari et al., 2020) questioned whether the beneficial outcomes resulting from positive displayed emotions by a chatbot would also be expectable in critical service encounters such as recovery situations. Additional concerns in this regard came from the emojis, in combination with wit, needed to display positive emotions. However, the results show that displaying positive emotions elicits favorable customer responses even in critical recovery encounters. The presumed moderation effect caused by an unsuccessful service recovery and the resulting limited possibility of empathizing with the chatbot could not be confirmed. This was determined by the fact that the effect of positive displayed emotions on empathy was not significantly stronger in a successful recovery than in an unsuccessful one. By changing the underlying service scenario in the aforementioned study, the study also contributed to generalizing the previously gained insights. In general, on the one hand, the multimethod approach to measure emotional reactions by the participants could rule out methodological limitations. On the other hand, the change of the used method to design participants' interaction with the chatbot and changes in the underlying scenario from a routine booking of a hotel room to the handling of a service recovery has shown the following: the effects around emotional contagion and empathy, with their positive effects on the satisfaction with service and recovery efforts, are valid in different service encounters.

Implications and Limitations 7

7.1 Theoretical Implications

Various research studies with different priorities have demonstrated the relevance of emotions for marketing in general and service encounters in particular (e.g., Grandey et al., 2005; Kenning et al., 2005; McClure et al., 2004; Pugh, 2001). An especially relevant phenomenon in the latter respect is emotional contagion (Hatfield et al., 1992). With the spread of chatbots in service encounters increasingly replacing frontline employees (Larivière et al., 2017; Marinova et al., 2017; Mozafari et al., 2020; Verhagen et al., 2014), the inevitable question has been raised about whether the transmission process of emotions and the associated positive outcomes, from a service perspective, will be lost. An answer to this question was the core objective of this thesis. In addition, the six consecutively conducted studies focused on empathy as a more cognitive and more conscious counterpart to emotional contagion (e.g., Decety & Lamm, 2006; Prochazkova & Kret, 2017). With this, this thesis adds to extant findings of different research streams. The most central finding here is that, in the context of chatbots, it is not so much a question of whether they can feel emotions but rather whether they create the feeling that they could feel emotions and present them authentically. Displaying emotions may cause social responses toward chatbots that are known from human-to-human interactions (e.g., emotional contagion). Thus, it appears that emotions in the context of AI and chatbots should be discussed primarily on the impact level rather than the level of origin. This largely corresponds to the opinion suggested by supporters of the computing perspective (e.g., Minsky, 2006; Picard, 1995). These vital results will be reviewed, and their implications presented in the further course.

K. Prinz, *The Smiling Chatbot*,
https://doi.org/10.1007/978-3-658-40028-6_7

Emotional contagion

Regarding emotional contagion, the present thesis contributes to the two research streams on emotional contagion and on conversational agents by bringing both together. By confirming that it is possible for emotional contagion to occur from a chatbot to a human being, this thesis joins the series of existing research on, for example, robots and emotional contagion (e.g., Tielman et al., 2014; Tsai et al., 2012) and complements them in important respects, especially concerning the service context in which the present thesis took place. Particularly noteworthy is the extension of the results away from a physical representation of an artificial agent to a purely text-based representation. In line with the CASA paradigm (Reeves & Nass, 1996), the results confirm that people unconsciously show behavioral patterns and responses in their interactions with chatbots, which are commonly associated with human-to-human interactions. This clearly indicates that humans treat chatbots as social interaction partners.

More importantly than the mere occurrence of emotional contagion, the way this contagion happens must be pointed out. Furthermore, the beneficial outcomes in service encounters are noteworthy when a contagion with positive emotions occurs. The results indicate that emotional contagion triggered by a chatbot is based on similar mechanisms as would be the case in a contagion between two humans. This applies to the transmission itself but also the outcomes. Regarding the process of emotional contagion, the results deliver strong support for the assumption that emotional contagion triggered by a chatbot also occurs via a three-step process (Barsade, 2002). Using emojis combined with wit as substitutes for facially expressed emotions, the results indicated that those participants who were exposed to those emotions showed a significant change in their facial valence toward more positivity being expressed. The subsequent consecutively conducted studies showed participants self-reporting significantly more positive affect when exposed to a chatbot displaying positive emotions. In this way, this thesis, at least in parts, contradicts Derks et al. (2008, p. 779), who stated in their work: "Mimicry, however, is impossible via text-based CMC [computer-mediated communication], and cannot be fulfilled by the use of emoticons." Instead, the extant research (Lohmann et al., 2017; Smith & Rose, 2020), as well as the results of the present thesis, show clear support in favor of the assumption that computer-mediated communication does not inhibit mimicry. With these results, this thesis closes a critical gap in research regarding conversational agents and chatbots in particular, as the known research projects addressed this question based on a sample that either consisted only of children (Tielman et al., 2014) or was conducted in a human-to-human context (Smith & Rose, 2020). Also, in terms of the outcomes of a transmission of positive emotions to a customer, the results from the interaction with a chatbot provide the same positive effects on encounter satisfaction as known from human-to-human research (Barger & Grandey, 2006; Pugh, 2001; Tsai & Huang, 2002). This provides central evidence for the validity of the affect-as-information theory (Clore et al., 2001).

In its entirety, the presented results confirm the CASA paradigm's existence (Reeves & Nass, 1996). Thus, the thesis is in line with previous approaches that could confirm the CASA paradigm in the context of chatbots (e.g., Ho et al., 2018). Vital new insights in this respect are provided as the validity of the paradigm is extended to the field of disembodied chatbots. In combination with the use of emojis and wit to substitute facially expressed emotions, the thesis confirms the application of social phenomena from interpersonal interactions to interactions with chatbots. This supports the trend in research that conversational agents and chatbots are increasingly being investigated from the perspective of social interaction partners. In this respect, the thesis is in line with the call made by various pioneers in the field of AI, in this context, for more consideration of human behavior and its patterns as a starting point for the development of AI technologies, rather than simply focusing on the completion of tasks (e.g., Beal & Winston, 2009; McCarthy, 2007; Nilsson, 2005). The thesis thus also corresponds to the assumption already expressed by van Doorn et al. (2017), who stated that in the future it will be increasingly important in interactions with technology that it is also fulfills social tasks and conforms to social rules in interactions.

With the presented results around emotional contagion from a chatbot to a human, the thesis provides insights not only in this context. On a higher level, theoretical conclusions can also be drawn, which are of considerable importance regarding the text-based expression of emotions. The commonly-held opinion that an exchange of social aspects such as emotions is largely lost through text-based communication channels (Byron, 2008; Huang et al., 2008; Kiesler et al., 1984; Walther & D'Addario, 2001) must be decisively contradicted at this point. On the one hand, these findings apply to the communication of people and conversational agents. On the other hand, it can be assumed that these findings are also valid in human-to-human communication. This topic will be taken up again and addressed in more detail at a later point.

With regard to the emotional contagion from a chatbot to a customer, however, it must be noted that the personality traits of a customer play an essential role. The results show that extraverted customers responded more strongly to the affective chatbot than introverted ones. This means that the contagion process was stronger for extraverted customers because they were more open to the emotions of the chatbot (Larsen & Ketelaar, 1989). Thus, the results provide evidence that extraverts are not only more likely to infect others with their emotions, as known from human-to-human research (e.g., Friedman & Riggio, 1981; Sullins, 1991), but, in a chatbot context, they are also more likely to catch a chatbot's emotions. Overall, the results show that the interaction with an affective chatbot is an encounter to be studied taking into consideration the personal traits of an individual. This thesis thus extends current knowledge that found customers preferring a chatbot's personality that is similar to their own (Jung et al., 2012) and that matches the business context (Jain et al., 2018). Thus, it becomes clear that personality traits in research on chatbots must be considered from both directions (i.e., personality of the chatbot and personality of the customer).

Empathy

The second essential stream of this thesis focuses on the feeling of empathy. Based on the definition of Rogers (1959), empathy in the present thesis was considered a cognitive counterpart to emotional contagion. To be precise, the focus was on empathy as the situational ability to put yourself into the emotional state of the chatbot. The essential difference between the two constructs can be found in the fact, that with empathy, a clear separation always remains between the affective state of the counterpart and one's own state. There is a mental projection of the other's emotions but no unconscious transmission. The basis for the ability to put oneself in the affective state of the chatbot was the theory around anthropomorphism, the conscious attribution of human characteristics to non-human entities (Epley et al., 2007). Looking at the feeling of empathy toward an artificial entity from the perspective of an interacting human has already been the subject of extant research approaches, particularly in the field of robotics (e.g., Paiva et al., 2004; Rosenthal-von der Pütten et al., 2013; Rosenthal-von der Pütten et al., 2014). These studies, however, were subject to a vital limitation, as they investigated empathy based on either inflicting pain on the robot or bullying it. Furthermore, robots are more likely to trigger anthropomorphic reactions due to their embodied nature (Duffy, 2003). With the results found, this thesis can confirm previous research and at the same time complement it with an underlying positive understanding of emotions and related empathy. Furthermore, the results show that an embodied nature is not necessary for anthropomorphism. Instead, the same effects can be elicited by human-like behavior.

In another respect, this thesis also provides a new approach in that it looks at the topic of empathy from a perspective, which has so far received little attention in service research, especially in a chatbot context. This approach is based on the fact that empathy is regarded in scientific research as the cognitive and conscious counterpart to emotional contagion (e.g., Decety & Lamm, 2006; Prochazkova & Kret, 2017) (see Section 3.2.3 for a discussion). Based on this view of research, empathy was included in the proposed research model as a second mediation path from the representation of positive emotions to encounter satisfaction. This mediation effect has been confirmed in several independent studies. It must also be highlighted that the investigated paths through customer positive affect and empathy fully mediated the relationship between a chatbot's positive displayed emotions and encounter satisfaction. Both processes can, as such, fully explain the effects caused by the displayed emotions. Thus, the thesis essentially contributes to the existing literature by showing that understanding the emotional states in a service transaction is not a one-sided business. Instead, it seems to depend on a mutual understanding of the respective affective state of the counterpart. Related research is relatively scarce. One of the few studies was conducted by Wieseke et al. (2012). They found that empathy from the client's perspective is a central driving force for the development and strength of the client-employee relationship. Gremler and Gwinner (2000) see the attention toward the affective state and, thus, in a broader sense, the customer's adoption of the employee's perspective, as a kind of evaluative approach to the employee's abilities. This means that, while research around emotional labor assumes that the evaluation process of authenticity of the emotions shown is an unconscious process (Hennig-Thurau et al., 2006), the approach mentioned above seems to assume a more conscious evaluation basis. In the context of the present research, it can be speculated that the feeling of empathy toward the chatbot also represents an evaluation of the appropriateness of the emotions displayed. For the research on chatbots, these results underline the fact that service encounters do not exclusively build on the fulfillment of functional requirements but that, in analogy to human-to-human interactions, social aspects also influence the evaluation of the service.

Human-like cues and graphical representations of chatbots

In addition, the present thesis is one of the first to investigate the effect of the transmission of emotions between a chatbot and a person with the theory around human-like cues and the impact on anthropomorphizing chatbots. As previously explained, research on emotional contagion in the context of AI is not new. Initial approaches have investigated this with a strong focus on robotics (Tielman et al., 2014). However, what is new is the strong reduction of social and emotional cues to text-based communication and representing AI technologies as a chatbot. This is important as robots typically cause strong anthropomorphic and social reactions due to their physical appearance (Duffy, 2003). Regarding causing anthropomorphic and social reactions toward chatbots, the presented results could show that a chatbot is able to change the affective state of the customer by displaying positive emotions. Furthermore, this process takes place not only on a subconscious level (primitive emotional contagion), but there is also the process of mentally representing the perceived emotions (empathy). With these findings, it could be shown in a series of consecutive studies that people ascribe an own affective state to the chatbot. The extant literature suggests that anthropomorphizing chatbots can be supported by using informal language, a human name, and the general reference of the communication behavior to human-based behavior (Araujo, 2018). The present thesis, on the one hand, confirms these results and, on the other hand, extends them by including emojis and wit as substitutes for facially displayed emotions during face-to-face interactions. An explanation for these results can be found in the fact that emotions are an essential part of human interaction (Felbo et al., 2017). Understandably, displaying typical human behavioral patterns serves as a human-like cue in interactions between humans and chatbots.

In this regard, it was investigated how graphical and behavioral human-like cues behave in combination. The results showed that displaying an avatar, meaning a graphical representation of the chatbot, can have different effects on the investigated affective reactions. The present thesis complements and partly contradicts extant research on anthropomorphism and designing chatbots at a central point. While previous research assumed that it is primarily a physical or graphical human-like representation of an agent that causes anthropomorphic behavior toward the agent, for example, a chatbot (e.g., Corti & Gillespie, 2016; Lee et al., 2010; von der Pütten et al., 2010), the results of this thesis show that this generalization is not valid. The presented results show that portraying the chatbot with an avatar featuring anthropomorphic features may improve the effects caused by displaying positive emotions, compared to a strongly computer-like avatar featuring no human-like attributes. However, the results showed no improvement caused by the human-like avatar compared to the situations where the chatbot was not represented by any graphical representation. These effects may be due to a decreasing marginal benefit of adding more human-like cues. This means that if enough cues are present to indicate human likeness, such as human-like communication behavior, an additional cue through a graphical representation is not necessary to strengthen the effect. However, it is possible to disturb the effects caused by the communication behavior by displaying cues that point away from human likeness, such as an avatar that highlights the chatbot's technical attributes and indicates no human likeness at all. The non-existent enhancing effect of a human-like avatar compared to no avatar at all may, in this regard, be caused because the customers have developed a mental image of the chatbot, when no graphic representation was available. This representation may be disturbed by the avatar when it is presented. However, no final explanation for the reasons can be provided at this point. Therefore, future research should address this point to further illuminate the effects of a chatbot's graphical representation in combination with the behavioral aspects.

Application scenarios for chatbots

Several existing studies that reported reservations toward chatbots in certain service situations prompted the investigation of the affective delivery of a chatbot in the context of a service recovery. For example, Mozafari et al. (2020) have shown that trust issues can arise in critical chatbot-handled service encounters. In addition, other studies have reported negative effects of the disclosure that customers are interacting with a chatbot on acceptance (Murgia et al., 2016) or persuasion (Shi et al., 2020). The results of this thesis speak a partially different language. In a recovery context, it could be shown that customers exhibited unconscious social reactions toward the chatbot. In addition, they anthropomorphized the chatbot, despite the critical service situation, and expressed more empathy. This creates some tension between the results of existing research and this thesis. The central question here is why the study of the present thesis concludes that the emotional performance of a chatbot, even in critical service encounters, leads to reactions by the customers that do not suggest that there are reservations about chatbots.

It is assumed that the reservations mentioned by customers in previous studies may have been caused by the fact that chatbots were not able to adequately meet social requirements (Huang & Rust, 2018; Wallis & Norling, 2005). This assumption is supported by the fact that humans are social beings and emotions play a vital role in interpersonal interaction (Ekman, 2000; Ekman et al., 1980; Myers, 1989). It has been shown in various studies and contexts that service encounters are also evaluated based not only on functional aspects, but that social components are also important (Oliver, 1993; Smith & Bolton, 2002; Westbrook & Oliver, 1991). Based on the CASA paradigm, it was pointed out that the unconscious social reactions of people toward artificial entities are due to learned patterns from a time when the only beings who exhibited social behavior were humans (Reeves & Nass, 1996). In this respect, it is not surprising that the same mechanisms of judgment are found in human-to-chatbot interactions. For research around chatbots, this means that a distorted picture is created when research is only engaged in investigating purely functional aspects that may drive customer satisfaction. The lack of performance of the chatbot on the social level of a service situation can then become problematic, which ultimately leads to aversion on the part of customers.

What could be observed, however, was that customers' reactions to the positive emotions of the chatbot were weaker in recovery situations than in previous routine encounters. Nevertheless, this can also be used as a support for the above-mentioned point of view (i.e., that chatbots need to display appropriate emotions according to the context). In recovery situations, in which an unfavorable situation has arisen for the customer, it is possibly less relevant to be exposed to positive emotions. Instead, customers in such situations expect other types of behavior, such as courtesy (Hocutt et al., 2006). This may explain why emotional contagion was weaker in this context. However, this supports the view that chatbots must also act on a social level in critical encounters. The results of the present thesis thus support the statement by Wallis and Norling (2005, p. 35) who state: "[…] to be an acceptable interface, a conversational agent must be accepted as a social actor, must display social intelligence and must participate in the social hierarchy." It is therefore crucial that in future research on chatbots, the influence of social behaviors is considered to a much greater extent.

Ways to display positive emotions

The research stream on computer-mediated communication was part of the basis for the conceptual development of the thesis. Since this research stream focuses on human-to-human interactions, the results of the present thesis can also contribute to the development of a more profound understanding in this respect. Of particular importance is the communication of emotions. In this regard, various studies have indicated that the use of emojis can lead to the occurrence of unfavorable customer reactions from the perspective of service providers (Duijst, 2017; Thies et al., 2017). These reactions were one of the reasons why the applicability of the chatbot in a critical recovery context was evaluated in the sixth study of this thesis. The results showed that even in such situations the use of emojis, in combination with wit, to express positive emotions from a chatbot is associated with beneficial service outcomes. These divergent results raise the question of why other studies report problems with the use of emojis, and the present thesis comes to different conclusions. Two possible explanations come to mind to explain the different results.

First, the results of the thesis indicate that, for the communication of emotions in computer-mediated communication, comparable mechanisms apply as they are known from face-to-face communication. In this respect, the authenticity of the perceived emotions is particularly noteworthy (Hennig-Thurau et al., 2006). The pre-tests conducted have shown that people are able to recognize incongruities between the statement made by the emojis used and the surrounding text. In the present case, the representation of positive emotions using emojis in combination with neutrally-worded text resulted in the emotions being perceived as inauthentic. Only a congruent wording of the text matching the statement of the emojis led to a perception of authenticity. It is possible that previous studies had not taken this into account, leading to a negative perception of the use of emojis. In this case, however, it would not have been the emojis that led to the unfavorable evaluations, but the incongruity.

Second, research on text-based communication of emotions via computer-mediated channels still faces many unresolved questions. This is partly because emojis in particular have only gained massive popularity in recent years (Felbo et al., 2017). One must consider that the expression of emotions in face-to-face communication is an unconscious process, compared to communicating emotions in a text-based way (Walther & D'Addario, 2001). Here, it can be assumed that the use of emojis has become a habit for most people. From personal experience, it can be stated that, although the use of emojis does not follow an unconscious schema, their increasingly frequent use is less influenced by conscious deliberation processes. From this, it can be concluded that with regard to the previously mentioned studies, which were able to show negative effects concerning emoji use, the question of the time when the reactions were evaluated must be asked. If emojis were still a relatively new phenomenon at the time when the studies were conducted, it cannot be ruled out that the time difference between the present thesis and the studies mentioned is responsible for the different effects, as emojis today are much more prevalent in everyday communication.

7.2 Practical Implications

An essential part of the introduction was the statement that increasing numbers of frontline employees are being replaced by chatbots (Larivière et al., 2017; Marinova et al., 2017; Mozafari et al., 2020; Verhagen et al., 2014) and that this raises the inevitable question of to what degree this exchange process will cause emotional components of transactions to get lost. Of special interest in the present thesis was emotional contagion. With the confirmation that the contagion process is possible from chatbots to humans, the thesis delivers important results for managers planning to install chatbots in customer service or developers of such technologies, which will be discussed in more detail below.

Affective chatbots as social actors and their service relevance

Looking at the motives for the use of AI in general, and chatbots in particular, it is noticeable that many authors highlight functional aspects such as permanent accessibility (Schneider, 2017; Wünderlich & Paluch, 2017), consistent service quality (Behera, 2016; Schanke et al., 2021), or saving potentials regarding labor costs (Li et al., 2017). However, these approaches obviously neglect the key strengths of chatbots. This thesis provides clear evidence that the interaction between chatbots and humans follows the rules that would be used for human-to-human interaction and, therefore, positive outcomes showing positive emotions can be expected. For the practical use of a chatbot, this means that the same standards must be applied when preparing chatbots (i.e., the programming) for the tasks to be performed. While one would never think of training employees in customer contact in only the technical aspects, the focus on functional advantages of chatbots seems to suggest exactly that. The same efforts that managers put into training employees in social skills should therefore be put into designing chatbots. Besides typical essential aspects like correct spelling and grammar (Morrissey & Kirakowski, 2013), which are among the basic characteristics, less obvious characteristics, like emotional behavior, have to also be considered. With the progressive replacement of employees by chatbots (Larivière et al., 2017; Marinova et al., 2017; Mozafari et al., 2020; Verhagen et al., 2014), a social interaction partner is lost if chatbots are trained only on a functional level. It seems that customers do not want to engage in encounters without social exchange; hence, chatbots must fill this gap. In this respect, it should be noted that the path to a correct presentation of emotions is not as trivial as it may appear at first glance. That is, even in a chatbot context, the perception of emotions as authentic plays a crucial role. This corresponds to the typical situation in human-to-human interactions, where it is well documented that authenticity plays a critical role in emotional contagion and beyond for the emergence of its positive outcomes (Hennig-Thurau et al., 2006). For example, it is also assumed that employees can learn to show an authentic smile even in situations in which the emotions presented do not correspond to those felt (surface acting) (Ekman & Friesen, 1982). It would be fatal if managers left the potential unused in this respect and continued to view chatbots only from the perspective of the task performers.

Susceptible customer segments for interactions with affective chatbots
Concerning emotional contagion between a chatbot and a customer, the results also hold important implications for customer relationship management. It turns out that customers who are characterized by a high degree of extraversion react more positively to the positive emotions of the chatbot. This means that the contagion process is stronger with these customers. More introverted customers may therefore react either less strongly or even negatively to the positive emotions of a chatbot. For companies that intend to use affective chatbots, this means that they should know the prevailing personality traits of their customer segments. It is important to keep in mind that even comparatively stable personality traits can be influenced by situational factors. This means that it is not impossible that introverted customers could exhibit extraverted behavior (i.e., state extraversion) in certain situations (Matthews, 1997). This state extraversion can be stimulated, for example, by giving customers as much control over the course of the conversations with the chatbot as possible (i.e., acting extraverted) (McNiel & Fleeson, 2006). This allows companies to stimulate beneficial effects, for example, through design features such as a chatbot that lets customers get the feeling they have control over the conversation and the situation.

This can be achieved, for example, by providing regular feedback on the current and next steps taken by the chatbot to enhance customer participation (Auh, Menguc, Katsikeas, & Jung, 2019). In addition, the chatbot can repeatedly pose questions in order to build a common understanding with the customer.

Approaches to express emotions in text-based interaction

An important implication from the presented results concerns the use of emojis as a substitute for facially displayed emotions. In the course of the thesis, a combination of the 😁-emoji and wit (i.e., informal language) specifically emerged as a positively and authentically perceived way to express positive emotions. However, it is important to note that there are crucial aspects that need consideration in this respect. As a conclusion from the empirical studies carried out, it must be noted that the use of emojis is by no means to be taken lightly. This means that the results obtained from the pre-studies conducted show the complex conception of the texts used, and they emphasize that people can recognize dissonances between the text-based statements and the statements of the added emojis. To generate the beneficial effects, it is necessary to align the two aspects. It is noticeable that, although the question around the text-based communication of emotions has been relevant for decades, especially since the emergence of e-mail communication (Rezabek & Cochenour, 1998), research on this topic is still comparatively rare. Thus, the underlying mechanisms are not yet fully understood. In this context, it must also be noted that, although the results of the present thesis are in line with related research approaches from human-to-human research (e.g., Smith & Rose, 2020), it has yet to be clarified to what extent a transfer of the context is possible. The results of Duijst (2017), for example, show that, under certain circumstances, customers may show negative reactions when emojis are used. This means that the extant research has shown that, from the customer's perspective, in critical situations an unfavorable perception of the use of emojis can occur. This contrasts with the finding in the present thesis that apparently no unfavorable perceptions by customers have arisen, even in the critical recovery situation. Therefore, it remains unexplained what influence the business context has on the use of emojis. It is therefore recommended proceeding with caution when using emojis as a substitute for facially expressed emotions, and to regularly incorporate customer feedback regarding their perception into the design. A major challenge in this conception may be interfering effects as emojis are typically represented differently across device platforms, which, even if the design process has been carefully conducted, harbors the risk of misinterpretations (Miller et al., 2016). From the company's point of view, this makes it even more important to engage in a regular exchange with customers. In particular, it must be ensured that feedback is obtained across all platforms to check for consistent interpretations.

With regard to the text-based expression of emotions, it must also be emphasized that emojis are only half the solution. It is not recommended to rely purely on the obviously simple implementation via emojis. Instead, it is important to adapt the wording of the text to the intended message of the emojis, for example by using wit.

The general basis and starting point for the investigation were the previous research results, which originate from the field of computer-mediated communication between humans (e.g., Lohmann et al., 2017; Smith & Rose, 2020). Although the present thesis deals with the text-based communication of a chatbot, the results also provide central implications for human-to-human communication via text-based computer-mediated channels. In this context, it becomes apparent that the text-based expression of emotions should be given more attention. For managers, the procedure of this thesis can serve as starting points for enriching computer-mediated communication during service encounters with social information, which may significantly influence the evaluation of the service. This is important because it can be expected that the already high relevance of messenger applications in service (Fowler, 2018) will continue to increase in the future.

Graphical representations of chatbots and their effects on perceiving chatbots as social actors

The results of the thesis regarding the graphical design of chatbots are also of high practical relevance, which arises from the many different approaches that prevail in practice about the design of avatars. While some companies use their corporate logo as an avatar (e.g., HelloFresh), others resort to fictional characters such as cartoon-like representations of humans (e.g., Lufthansa) or robots (e.g., Deutsche Bahn). From the results, the conclusion can be drawn that the design of the avatar can have a significant impact on the effects of the behavioral design of a chatbot. The effect emerged from the fact that adding a human-like avatar could not enhance the studied affective responses of customers to the positive emotions of the chatbot. This was true at least when the effects were compared to a situation without an avatar. On the contrary, the addition of an avatar, which strongly emphasized the technical nature of the chatbot, clearly inhibited the contagion of positive emotions between the chatbot and the customers. The human-like cues (Blut et al., 2021), which customers take as a point of reference to guide their behavior toward the chatbot, serve as an explanatory approach (Epley et al., 2007). These human-like cues seem to have a decreasing marginal benefit. This means that the addition of further cues can hardly strengthen the effects of the previous ones. At the same time, however, it must be ensured that the cues shown are congruent with each other. This is indicated by the finding that the affective customer reactions to the comparatively complex design of the chatbot's behavior could be prevented by a simple graphical representation that contradicted this. Accordingly, when developing chatbots, care must be taken to ensure congruence between behavior and appearance to prevent unfavorable effects. In a nutshell, this means, that, if the intention is to design a chatbot to be human-like in terms of its behavior, this should be supported by a corresponding avatar. If it is not possible to place such an avatar, it is recommended to show no avatar at all rather than choose an avatar that contradicts the human-like behavior.

Use cases for chatbots and their beneficial affective delivery

A key strength of chatbots is their ability to consistently deliver a certain level of service quality (Behera, 2016; Schanke et al., 2021). In addition, chatbots are neither susceptible to catching the occasionally prevalent negative emotions of customers (Du et al., 2011) nor are they at risk of being influenced by busyness in their affective delivery (Pugh, 2001). A particularly noteworthy advantage from the customer's perspective is the significantly higher efficiency in processing service requests. While typical attempts to contact companies usually involve waiting which harms customer satisfaction (del Río-Lanza et al., 2009; Maxham III & Netemeyer, 2002), chatbots enable much faster and wait-free processing. Since many customers do not make a complaint after a failure because they believe that the service provider will not respond (Singh, 1990), the use of chatbots can prevent the risk of missing a great opportunity to turn dissatisfied customers into satisfied ones. This makes chatbots excellently suited for use in service recoveries. This is crucial since a central strength of the present results is that they show that emotional contagion is also relevant in the course of negative situations, in the sense of mitigating negative emotions. Because employees are vulnerable to customers' negative emotions in the aftermath of a service failure (Dallimore et al., 2007), and chatbots are not, this is important. Instead, a strength of chatbots is their robustness against catching customers' negative emotions. In combination with emotional contagion resulting from the expression of positive emotions, favorable outcomes can be expected even during recovery situations. It should be noted that the positive effects occur even in situations when the chatbot is not able to provide an immediate solution to the problem encountered but is forced to refer the customer to an employee. In sum, this shows that chatbots are an ideal first point of contact for customers regardless of whether the recovery is successful (i.e., the chatbot can directly solve the problem that has arisen) or unsuccessful (i.e., the chatbot has to refer the customer to the customer service). This can mean a reduction in the workload of employees if the chatbot has been able to independently resolve initial concerns beforehand. For employees, this means more available resources to develop solutions for the remaining, more complex concerns and to satisfy customers (Sawhney, 2016). In addition, the chatbot has probably already been able to reduce customers' negative affect through displaying positive emotions, so that the downstream interactions between employees and customers are less susceptible to emotional escalation.

7.3 Limitations and Directions for Future Research

The research activities of this thesis are subject to several limitations, which will be addressed below and thus pave the way for future research approaches. First, the scenario-based approach offered the potential to isolate and exclude disturbing factors (Pascual-Leone et al., 2016), so that basic results on the topic could be obtained. The weaknesses resulting from using video-based stimuli were eliminated due to the multimethod approach undertaken by this thesis. In line with existing results (e.g., Rosenthal-von der Pütten et al., 2013), future research should therefore be encouraged to use video-based stimuli that require subjects to mentally project into, for example, the perspective of a customer. This is shown by the results, which are constant across different methods (i.e., Wizard of Oz method, videos, real chatbot). Nevertheless, limitations remain concerning the scenarios used, as this thesis is based exclusively on scenarios taken from the hospitality industry. This decision was encouraged because this industry has previously been featured in research on emotional contagion between humans (e.g., Grandey et al., 2005). Even though the investigated effects remain stable across a booking scenario and a check-in-related recovery situation, the question remains unanswered as to the extent to which the results may be readily transferred to other business contexts (Schanke et al., 2021). It has been shown, for example, that in human-to-human interactions, the expression of emotions elicits stronger positive effects when the encounters are characterized by communal norms (i.e., situations intended to give benefits to the other person) (Li et al., 2019). By contrast, these effects are weaker when exchange norms (i.e., situations based on getting something back) are more prevalent. In addition, results of different research approaches could show the negative perception of emojis in business communication (e.g., Duijst, 2017; Lebovits, 2015; Li et al., 2019; Thies et al., 2017). At this point, therefore, further research is needed to shed more light on text-based emotional expression and thus expand or limit the generalizability of the results obtained so far.

Second, the primary goal of this thesis was to check whether emotional contagion from a chatbot to an interacting customer takes place if the chatbot displays positive emotions. In addition, the goal was to investigate the ability of a customer to put himself/herself in the affective state of the chatbot. Showing positive emotions is typically associated with positive service encounters (Groth et al., 2009; Yoo & Arnold, 2016). However, not only positive situations (i.e., routine encounters) were used in the presented studies. Instead, service recoveries were also investigated. This was considered important, as service encounters are inevitably associated with service failures (La & Choi, 2019). Furthermore, recovery situations may also be caused because of, for example, a defective product. As such, recovery situations are usually infected with negative emotions on the part of the customer (Du et al., 2011). Previous studies have shown that emotional contagion can be a two-way process and that in failure or recovery situations, it is possible for a contagion of negative emotions to affect the employee (Dallimore et al., 2007). However, results from Du et al. (2011) show that positive displayed emotions by an employee can mitigate a negatively valenced affective state on the part of the customer. In this regard, it was considered important to investigate the role of an affective delivery provided by a chatbot in such a context, as it is a crucial strength of chatbots that they are able to provide a constant level of service quality independent of disturbing factors such as bad moods (Schanke et al., 2021). This makes them well suited to use in recovery situations, as they can resist the tendency of human employees to catch the negative emotions of customers that are sometimes present. This assumption was supported by the results that emotional contagion from a chatbot to a human is also possible in recovery situations. However, these results are contrasted with findings from earlier research approaches that raise questions. For example, Mozafari et al. (2020) found that, in a critical service encounter, revealing that customers were interacting with a chatbot had a negative effect on trust. This supported earlier findings from Luo et al. (2019), who found that the disclosure of a chatbot led to significantly worse sales performance. From these results, it can be concluded that customers exhibit reservations toward chatbots. In this respect, the sometimes limited performance of conversational agents in general, and chatbots in particular (Ho et al., 2018; Skjuve et al., 2019), may play a central role. Nevertheless, it cannot be excluded that there are other, possibly psychological, reasons for this. These existing results raise the question of the extent to which the use of chatbots is practical in critical situations, such as recovery situations. On the one hand, there are the positive outcomes resulting from the affective delivery of a chatbot. On the other hand, however, there are apparently also expressed reservations toward chatbots. At this point, more research is urgently needed that deals with boundary conditions regarding the use of chatbots. Only in this way can a better understanding be developed of the situations in which the use of chatbots is recommended and those in which it is not advisable.

Third, to "teach" chatbots how to express adequate emotions, an in-depth understanding of computer-mediated communication, particularly the communication of emotions between humans, is necessary. Although this topic started to be investigated with the triumphal march of e-mails in the 1990s, there is still no sound scientific understanding of this complex topic. The complexity is demonstrated by looking at the results of extant research. For example, Smith and Rose (2020) used the 😊-emoji ("smiling face with smiling eyes" (Emojipedia, n.d.-c)) for their research in an American-English-speaking context. In a pre-test conducted for the present thesis, which was conducted in a German-speaking context, this emoji led to significantly less perceived authenticity. It is noteworthy that both studies were conducted in a business context. In the end, it was the 😀-emoji, in combination with parts of informal language (wit), that met the requirements for a positive and authentic perception in the present research approach. Thus, the thesis confirms the approach by Lohmann et al. (2017), who also worked with a German-speaking sample and also used the 😀-emoji. In the run-up to each test of the hypotheses, a t-test showed that the perception of positivity between the two scenarios was already statistically different. Nevertheless, in both experimental groups, several subjects were found who classified the emotions shown as extraordinarily negative, although assigned to the manipulated group (i.e., when positive emotions were shown). The same effect in the opposite direction was observed in the control group. In the context of an experimental design, it is part of the standard methodology to perform a treatment check and to remove test persons who indicated to not have recognized the experimental treatment (Sigall & Mills, 1998). Nevertheless, the question remains as to why, with regard to such a fundamental question as the classification of emotions as neutral or positive, the evaluations apparently diverge. This reveals that more research is needed to gain a better understanding of how emotions can be expressed in computer-mediated communication.

Fourth, the addition of the recovery scenario and the associated replication of the previous procedure in a different context in *Study 6* led to a major contribution regarding the generalizability of the results around emotional contagion and empathy. In this context, it could be shown that the affect-as-information theory, in combination with the change in the customer's affective state through the affective delivery of the chatbot, also leads to beneficial service outcomes in recovery situations. Various research approaches have shown that the aspect of justice is a central factor in the course of recovery encounters (e.g., La & Choi, 2019). Existing research has comparatively often investigated the effect of the perception of justice on the affective state of customers. Of special interest in the context of employees' affective delivery may be interactional justice. Interactional justice refers to how the recovery was carried out (Lee et al., 2011). As *Study 6* did not consider justice in the model as it was not part of the investigated question, no assessment can be given as to whether or how the expression of positive emotions and justice are related in a chatbot context. This is a question that future research approaches should address by bringing together the different research streams to a greater extent.

Fifth, the fourth study examined the moderating effects of the personality traits of extraversion and openness to experience. These two traits belong to the Big Five. The results of *Study 4* showed that, in line with the hypothesis, extraversion positively moderated the positive effect of positive displayed emotions of a chatbot on customer positive affect. For reasons of complexity, the remaining three personality traits (agreeableness, conscientiousness, and neuroticism) were not included as control variables in the model. For this reason, no statement can be made as to whether there are further effects that can be attributed to different personality traits. In order to be able to provide customer relationship management with further information on the customer segments in which the use of affective chatbots is advisable or not advisable, future studies should start at this point and close the gap.

Finally, the results of the conducted studies clearly show that an affective chatbot can trigger affective customer responses. In addition to the question of whether a chatbot is generally able to cause these effects, there is also the question of how its performance compares to that of a human employee. Furthermore, in the presented studies, the chatbot was always able to communicate at a comparatively high level. Therefore, the question can also be raised of what influence the chatbot's communication skills have on the existence and the magnitude of the investigated affective responses. This is especially important in comparison to the mentioned human performance. From these two aspects, it is recommended that future research approaches should examine how the performance of an affective chatbot compares to that of a human. In addition, it should also be evaluated whether the affective responses of the customers depend on the communication skills of the chatbot.

Overall, the conclusion must be drawn that the question around the representation of emotions by chatbots and human reactions still offers many unaddressed questions. However, the essential point is that chatbots provide much more than the purely functional processing of underlying service encounters. Instead, they have the potential to conform to social rules and fulfill social needs of customers through emotional components in service encounters. Although the results of this thesis provide clear guidance for the design of chatbots, there are still several open questions that make this topic a fertile field for further research.

7.4 Management Summary

This thesis has examined the aspect of customers' affective responses to positive displayed emotions by chatbots. The results obtained provide clear evidence that positive displayed emotions of a chatbot can trigger emotional contagion and the feeling of empathy toward chatbots, and both affective customer reactions positively influence the level of satisfaction with the service experienced. On the one hand, the series of studies clearly shows that customer satisfaction with chatbot-handled service encounters is also largely determined by emotional components. On the other hand, this means that a large part of the potential of chatbots is wasted if the focus is only on advantages such as accessibility or cost-efficiency. For this reason, it is strongly recommended that when using chatbots in service, care should be taken from the start of the development process to ensure that they can also meet customers on a social level. Particular attention should be paid to "teaching" chatbots to express appropriate emotions. For this, using emojis can be a good method to express these emotions in text-based communication. It should be noted, however, that emojis alone are not enough to be perceived as both positive and authentic. This requires subtle adjustments to the text, the message of which must match the intended message of the emojis. Furthermore, managers should know their own customers well. Not all customers are equally responsive to interactions with affective chatbots. Extraverted customers have emerged as a well-suited customer group in this respect.

The many different implementation approaches available show the high practical relevance of graphically representing chatbots through avatars. In this respect, the thesis provides important findings, some of which clearly contradict existing approaches. When implementing avatars, it should be borne in mind that it is congruent with the intended human-like design of the chatbot's behavior. This is not to reinforce the effects caused by the behavior, but to prevent those effects being inhibited. In short, this means that if a company intends to use a chatbot that is human-like in terms of its behavior, then the associated avatar should not contradict these human-like cues by, for example, highlighting the technical nature of the chatbot. It is also important to emphasize the potential for service providers concerning the handling of recoveries. On the functional level, the use of chatbots can prevent waiting time for customers. On the emotional level, the results show that positive displayed emotions by a chatbot can also trigger emotional contagion in such encounters. This can reduce a customer's negative affect even if the customer must be referred to an employee because the chatbot cannot provide an initial solution to remedy the failure. This once more highlights the potential associated with implementing affective chatbots in service. At this point, managers should be encouraged to increase the use of affective chatbots for the processing of service encounters and use the full potential that chatbots offer.

References

Allen, J. A., Pugh, S. D., Grandey, A. A., & Groth, M. (2010). Following display rules in good or bad faith?: Customer orientation as a moderator of the display rule-emotional labor relationship. *Human Performance, 23*(2), 101–115.

Alshamrani, R., & Ma, X. (2019). Deep learning. In L. Schintler & C. McNeely (Eds.), *Encyclopedia of Big Data.* Cham, Germany: Springer.

Amazon Web Services Inc. Was ist künstliche Intelligenz? [What is artificial intelligence?]. Retrieved March 7, 2019 from https://aws.amazon.com/de/machine-learning/what-is-ai/?nc1=f_ls

Anders, S., Heinzle, J., Weiskopf, N., Ethofer, T., & Haynes, J. D. (2011). Flow of affective information between communicating brains. *NeuroImage, 54*(1), 439–446.

Angga, P. A., Fachri, W. E., Elevanita, A., & Agushinta, R. D. (2015). Design of chatbot with 3D avatar, voice interface, and facial expression. In *2015 International Conference on Science in Information Technology (ICSITech)* (pp. 326–330). Yogyakarta, Indonesia: IEEE.

Araujo, T. (2018). Living up to the chatbot hype: The influence of anthropomorphic design cues and communicative agency framing on conversational agent and company perceptions. *Computers in Human Behavior, 85*, 183–189.

Arnold, R. D., & Wade, J. P. (2015). A definition of systems thinking: A systems approach. *Procedia Computer Science, 44*, 669–678.

Aronfreed, J. (1970). The socialization of altruistic and sympathetic behavior: Some theoretical and experimental analyses. In J. Macaulay & L. Berkowitz (Eds.), *Altruism and Helping Behavior* (pp. 103–126). New York: Academic Press.

Ashforth, B. E., & Humphrey, R. H. (1993). Emotional labor in service roles: The influence of identity. *Academy of Management Review, 18*(1), 88–115.

Astrid, M., Krämer, N. C., Gratch, J., & Kang, S. H. (2010). „It doesn't matter what you are!“ Explaining social effects of agents and avatars. *Computers in Human Behavior, 26*(6), 1641–1650.

Athanassopoulos, A., Gounaris, S., & Stathakopoulos, V. (2001). Behavioral responses to customer satisfaction: An empirical study. *European Journal of Marketing, 35*(6), 687–707.

Auh, S., Menguc, B., Katsikeas, C. S., & Jung, Y. S. (2019). When does customer participation matter? An empirical investigation of the role of customer empowerment in

K. Prinz, *The Smiling Chatbot*,
https://doi.org/10.1007/978-3-658-40028-6

the customer participation–performance link. *Journal of Marketing Research, 56*(6), 1012–1033.

Bagozzi, R. P., & Yi, Y. (1988). On the evaluation of structural equation models. *Journal of the Academy of Marketing Science, 16*(1), 74–94.

Barger, P. B., & Grandey, A. A. (2006). Service with a smile and encounter satisfaction: Emotional contagion and appraisal mechanisms. *Academy of Management Journal, 49*(6), 1229–1238.

Baron-Cohen, S., & Wheelwright, S. (2004). The empathy quotient: An investigation of adults with Asperger syndrome or high functioning autism, and normal sex differences. *Journal of Autism and Developmental Disorders, 34*(2), 163–175.

Baron, S., Harris, K., Elliott, D., Schoefer, K., & Ennew, C. (2005). The impact of perceived justice on consumers' emotional responses to service complaint experiences. *Journal of Services Marketing, 19*(5), 261–270.

Barr, C. L., & Kleck, R. E. (1995). Self-other perception of the intensity of facial expressions of emotion: Do we know what we show? *Journal of Personality and Social Psychology, 68*(4), 608–618.

Barrett-Lennard, G. T. (1998). *Carl Rogers' helping system: Journey and substance.* London: Sage.

Barsade, S. G. (2002). The ripple effect: Emotional contagion and its influence on group behavior. *Administrative Science Quarterly, 47*(4), 644–644.

Bartneck, C. (2002). Integrating the occ model of emotions in embodied characters. In *Workshop on Virtual Conversational Characters: Applications, Methods, and Research Challenges* (pp. 39–48). Melbourne, Australia.

Batson, C. D., Batson, J. G., Slingsby, J. K., Harrell, K. L., Peekna, H. M., & Todd, R. M. (1991). Empathic joy and the empathy-altruism hypothesis. *Journal of Personality and Social Psychology, 61*(3), 413–426.

Batson, C. D., Lishner, D. A., Cook, J., & Sawyer, S. (2005). Similarity and nurturance: Two possible sources of empathy for strangers. *Basic and Applied Social Psychology, 27*(1), 15–25.

Bavelas, J. B., Black, A., Lemery, C. R., & Mullett, J. (1986). „I show how you feel": Motor mimicry as a communicative act. *Journal of Personality and Social Psychology, 50*(2), 322–329.

Beal, J., & Winston, P. H. (2009). The new frontier of human-level artificial intelligence introduction. *IEEE Intelligent Systems, 24*(4), 21–23.

Behera, B. (2016). Chappie—a semi-automatic intelligent chatbot. Retrieved May 17, 2019 from https://www.cse.iitb.ac.in/~bibek/WriteUP_2016.pdf

Bellman, R. E. (1978). *An introduction to artificial intelligence: Can computers think?* San Francisco, USA: Boyd & Fraser Publishing Company.

Bente, G., Rüggenberg, S., Krämer, N. C., & Eschenburg, F. (2008). Avatar-mediated networking: Increasing social presence and interpersonal trust in net-based collaborations. *Human Communication Research, 34*(2), 287–318.

Bernieri, F. J. (1988). Coordinated movement and rapport in teacher-student interactions. *Journal of Nonverbal Behavior, 12*(2), 120–138.

Bickmore, T. W., & Cassell, J. (2005). Social dialogue with embodied conversational agents. In J. C. J. v. Kuppevelt, L. Dybkjær, & N. O. Bernsen (Eds.), *Advances in Natural Multimodal Dialogue Systems* (pp. 23–54). Dordrecht, The Netherlands: Springer.

Bijlstra, G., & Dotsch, R. (2011). FaceReader 4 emotion classification performance on images from the Radboud Faces Database. *Unpublished manuscript, Department of Social and Cultural Psychology, Radboud University Nijmegen, Nijmegen, The Netherlands.*

Biocca, F., & Harms, C. (2002). Defining and measuring social presence: Contribution to the networked minds theory and measure. In *Proceedings of PRESENCE* (pp. 7–36). Porto, Portugal: University of Fernando Pessoa Press.

Bitner, M. J., Brown, S. W., & Meuter, M. L. (2000). Technology infusion in service encounters. *Journal of the Academy of Marketing Science, 28*(1), 138–149.

Blut, M., Wang, C., Wünderlich, N. V., & Brock, C. (2021). Understanding anthropomorphism in service provision: A meta-analysis of physical robots, chatbots, and other AI. *Journal of the Academy of Marketing Science, 49*, 632–658.

Boiteux, M. (2019). Messenger at F8 2018. Retrieved May 14, 2019 from https://blog.messengerdevelopers.com/messenger-at-f8-2018-44010dc9d2ea

Borke, H. (1971). Interpersonal perception of young children: Egocentrism or empathy? *Developmental Psychology, 5*(2), 263–269.

Bowling, N. A., & Hammond, G. D. (2008). A meta-analytic examination of the construct validity of the Michigan organizational assessment questionnaire job satisfaction subscale. *Journal of Vocational Behavior, 73*(1), 63–77.

Broniarczyk, S. M., & Alba, J. W. (1994). The role of consumers' intuitions in inference making. *Journal of Consumer Research, 21*(3), 393–407.

Buckholtz, J. W., & Marois, R. (2012). The roots of modern justice: Cognitive and neural foundations of social norms and their enforcement. *Nature Neuroscience, 15*(5), 655–661.

Burke, M. J., Brief, A. P., George, J. M., Roberson, L., & Webster, J. (1989). Measuring affect at work: Confirmatory analyses of competing mood structures with conceptual linkage to cortical regulatory systems. *Journal of Personality and Social Psychology, 57*(6), 1091–1102.

Byron, K. (2008). Carrying too heavy a load? The communication and miscommunication of emotion by email. *Academy of Management Review, 33*(2), 309–327.

Carr, L., Iacoboni, M., Dubeau, M. C., Mazziotta, J. C., & Lenzi, G. L. (2003). Neural mechanisms of empathy in humans: A relay from neural systems for imitation to limbic areas. *Proceedings of the National Academy of Sciences, 100*(9), 5497–5502.

Cassell, J., & Thorisson, K. R. (1999). The power of a nod and a glance: Envelope vs. emotional feedback in animated conversational agents. *Applied Artificial Intelligence, 13*(4–5), 519–538.

Chakrabarti, C., & Luger, G. F. (2015). Artificial conversations for customer service chatter bots: Architecture, algorithms, and evaluation metrics. *Expert Systems with Applications, 42*(20), 6878–6897.

Chartrand, T. L., & Bargh, J. A. (1999). The chameleon effect: The perception-behavior link and social interaction. *Journal of Personality and Social Psychology, 76*(6), 893–910.

Chaves, A. P., & Gerosa, M. A. (2021). How should my chatbot interact? A survey on social characteristics in human-chatbot interaction design. *International Journal of Human-Computer Interaction, 37*(8), 729–758.

Chen, G., Ployhart, R. E., Thomas, H. C., Anderson, N., & Bliese, P. D. (2011). The power of momentum: A new model of dynamic relationships between job satisfaction change and turnover intentions. *Academy of Management Journal, 54*(1), 159–181.

Chin, W. W. (1998). The partial least squares approach to structural equation modeling. In G. A. Marcoulides (Ed.), *Modern Methods for Business Research* (pp. 295–336). Mahwah, USA: Lawrence Erlbaum Associates.

Chin, W. W. (2010). How to write up and report PLS analyses. In V. E. Vinzi, W. W. Chin, J. Henseler, & H. Wang (Eds.), *Handbook of Partial Least Squares: Concepts, Methods and Applications in Marketing and Related Fields* (pp. 655–690). Berlin, Germany: Springer.

Chin, W. W., & Dibbern, J. (2010). An introduction to a permutation based procedure for multi-group PLS analysis: Results of tests of differences on simulated data and a cross cultural analysis of the sourcing of information system services between Germany and the USA. In V. E. Vinzi, W. W. Chin, J. Henseler, & H. Wang (Eds.), *Handbook of Partial Least Squares* (pp. 171–193). Berlin, Germany: Springer.

Chin, W. W., Thatcher, J. B., Wright, R. T., & Steel, D. (2013). Controlling for common method variance in PLS analysis: the measured latent marker variable approach. In H. Abdi, W. W. Chin, V. E. Vnzi, G. Russolillo, & L. Trinchera (Eds.), *New perspectives in partial least squares and related methods* (pp. 231–239). New York, USA: Springer.

Cho, E. (2016). Making reliability reliable: A systematic approach to reliability coefficients. *Organizational Research Methods, 19*(4), 651–682.

Churches, O., Nicholls, M., Thiessen, M., Kohler, M., & Keage, H. (2014). Emoticons in mind: An event-related potential study. *Social Neuroscience, 9*(2), 196–202.

Clark, A. J. (2007). *Empathy in counseling and psychotherapy: Perspectives and practices.* London, UK: Lawrence Erlbaum.

Clore, G. L., Gasper, K., & Garvin, E. (2001). Affect as information. In J. P. Forgas (Ed.), *Handbook of affect and social cognition* (pp. 121–144). Mahwah, USA: Lawrence Erlbaum Associates.

Cohen, J. (1988). *Statistical power analysis for the behavioral sciences.* Hillsdale, USA: Lawrence Erlbaum Associates.

Colby, C. L., Mithas, S., & Parasuraman, A. (2016). *Service robots: How ready are consumers to adopt and what drives acceptance?* Paper presented at the 2016 Frontiers in Service Conference, Bergen, Norway.

Colby, K. M. (1975). *Artificial paranoia: A computer simulation of paranoid processes.* Oxford, UK: Pergamon Press.

Condon, W. S. (1982). Cultural microrhythms. In M. Davis (Ed.), *Interaction rhythms: Periodicity in communicative behavior* (pp. 53–76). New York, USA: Human Sciences Press.

Corti, K., & Gillespie, A. (2016). Co-constructing intersubjectivity with artificial conversational agents: People are more likely to initiate repairs of misunderstandings with agents represented as human. *Computers in Human Behavior, 58*, 431–442.

Cronbach, L. J., & Meehl, P. E. (1955). Construct validity in psychological tests. *Psychological Bulletin, 52*(4), 281–302.

Crutzen, R., Peters, G.-J. Y., Portugal, S. D., Fisser, E. M., & Grolleman, J. J. (2011). An artificially intelligent chat agent that answers adolescents' questions related to sex, drugs, and alcohol: An exploratory study. *Journal of Adolescent Health, 48*(5), 514–519.

Cunningham, M. R. (1979). Weather, mood, and helping behavior: Quasi experiments with the sunshine samaritan. *Journal of Personality and Social Psychology, 37*(11), 1947–1956.

Dahlbäck, N., Jönsson, A., & Ahrenberg, L. (1993). Wizard of Oz studies—why and how. *Knowledge-Based Systems, 6*(4), 258–266.

Dahm, M., & Dregger, A. (2019). Der Einsatz von künstlicher Intelligenz im HR: Die Wirkung und Förderung der Akzeptanz von KI-basierten Recruiting-Tools bei potenziellen Nutzern [The use of artificial intelligence in HR: The impact and driving adoption of AI-based recruiting tools among potential users.]. In B. Hermeier, T. Heupel, & S. Fichtner-Rosada (Eds.), *Arbeitswelten der Zukunft [Working environments of the future]* (pp. 249–271). Wiesbaden, Germany: Springer Fachmedien Wiesbaden.

Dallimore, K. S., Sparks, B. A., & Butcher, K. (2007). The influence of angry customer outbursts on service providers' facial displays & affective states. *Journal of Service Research, 10*(1), 78–91.

Danesi, M. (2016). *The semiotics of emoji: The rise of visual language in the age of the internet.* London, UK: Bloomsbury Publishing.

Daum, M., Holtgrewe, U., Schörpf, P., & Nocker, M. (2018). *Call- und Service-Center [Call and service centers].* Düsseldorf, Germany: Hans-Böckler-Stiftung.

Dautenhahn, K. (1999). Robots as social actors: Aurora and the case of autism. In *Proceedings Third Cognitive Technology Conference CT'99* (pp. 374–389). San Francisco, USA.

Davenport, T. H., & Ronanki, R. (2018). Artificial intelligence for the real world. *Harvard Business Review, 96*(1), 108–116.

Davis, M. H. (1983). Measuring individual differences in empathy: Evidence for a multidimensional approach. *Journal of Personality and Social Psychology, 44*(1), 113–126.

Davis, M. H. (1994). *Empathy: A social psychological approach.* Boulder, USA: Westview Press.

De Cicco, R., Silva, S. C., & Alparone, F. R. (2020). Millennials' attitude toward chatbots: An experimental study in a social relationship perspective. *International Journal of Retail & Distribution Management, 48*(11), 1213–1233.

De Cicco, R., Silva, S. C., & Alparone, F. R. (2021). „It's on its way": Chatbots applied for online food delivery services, social or task-oriented interaction style? *Journal of Foodservice Business Research, 24*(2), 140–164.

De Martino, B., Kumaran, D., Seymour, B., & Dolan, R. J. (2006). Frames, biases, and rational decision-making in the human brain. *Science, 313*(5787), 684–687.

Debauche, O., Mahmoudi, S., Mahmoudi, S. A., Manneback, P., Bindelle, J., & Lebeau, F. (2020). Edge computing and artificial intelligence for real-time poultry monitoring. *Procedia Computer Science, 175*, 534–541.

Decety, J., & Chaminade, T. (2003). Neural correlates of feeling sympathy. *Neuropsychologia, 41*(2), 127–138.

Decety, J., & Jackson, P. L. (2004). The functional architecture of human empathy. *Behavioral and Cognitive Neuroscience Reviews, 3*(2), 71–100.

Decety, J., & Jackson, P. L. (2006). A social-neuroscience perspective on empathy. *Current Directions in Psychological Science, 15*(2), 54–58.

Decety, J., & Lamm, C. (2006). Human empathy through the lens of social neuroscience. *The Scientific World Journal, 6*, 1146–1163.

del Río-Lanza, A. B., Vázquez-Casielles, R., & Díaz-Martín, A. M. (2009). Satisfaction with service recovery: Perceived justice and emotional responses. *Journal of Business Research, 62*(8), 775–781.

Deppe, M., Schwindt, W., Krämer, J., Kugel, H., Plassmann, H., Kenning, P., & Ringelstein, E. B. (2005). Evidence for a neural correlate of a framing effect: Bias-specific activity in the ventromedial prefrontal cortex during credibility judgments. *Brain Research Bulletin, 67*(5), 413–421.

Derks, D., Fischer, A. H., & Bos, A. E. (2008). The role of emotion in computer-mediated communication: A review. *Computers in Human Behavior, 24*(3), 766–785.

Devaraj, S., Easley, R. F., & Crant, J. M. (2008). Research note—how does personality matter? Relating the five-factor model to technology acceptance and use. *Information Systems Research, 19*(1), 93–105.

Dietvorst, B. J., Simmons, J. P., & Massey, C. (2015). Algorithm aversion: People erroneously avoid algorithms after seeing them err. *Journal of Experimental Psychology: General, 144*(1), 114–126.

Dimberg, U., Thunberg, M., & Grunedal, S. (2002). Facial reactions to emotional stimuli: Automatically controlled emotional responses. *Cognition & Emotion, 16*(4), 449–471.

Du, J., Fan, X., & Feng, T. (2011). Multiple emotional contagions in service encounters. *Journal of the Academy of Marketing Science, 39*(3), 449–466.

Duclos, S. E., Laird, J. D., Schneider, E., Sexter, M., Stern, L., & Van Lighten, O. (1989). Emotion-specific effects of facial expressions and postures on emotional experience. *Journal of Personality and Social Psychology, 57*(1), 100–108.

Duffy, B. R. (2003). Anthropomorphism and the social robot. *Robotics and Autonomous Systems, 42*(3–4), 177–190.

Duijst, D. (2017). Can we improve the user experience of chatbots with personalisation? *Master's thesis, University of Amsterdam, Amsterdam, The Netherlands.*

Dymond, R. F., Hughes, A. S., & Raabe, V. L. (1952). Measurable changes in empathy with age. *Journal of Consulting Psychology, 16*(3), 202–206.

Eisenberg, N., & Miller, P. A. (1987). The relation of empathy to prosocial and related behaviors. *Psychological Bulletin, 101*(1), 91–119.

Eisner, B., Rocktäschel, T., Augenstein, I., Bošnjak, M., & Riedel, S. (2016). emoji2vec: Learning emoji representations from their description. In *Proceedings of the 4th International Workshop on Natural Language Processing for Social Media (SocialNLP)* (pp. 48–54). Austin, USA: Association for Computational Linguistics.

Ekman, P. (2000). Basic emotions. In T. Dalgleish & M. Power (Eds.), *Handbook of cognition and emotion* (pp. 45–60). Hoboken, USA: John Wiley & Sons.

Ekman, P., & Davidson, R. J. (1993). Voluntary smiling changes regional brain activity. *Psychological Science, 4*(5), 342–345.

Ekman, P., Freisen, W. V., & Ancoli, S. (1980). Facial signs of emotional experience. *Journal of Personality and Social Psychology, 39*(6), 1125–1134.

Ekman, P., & Friesen, W. V. (1982). Felt, false, and miserable smiles. *Journal of Nonverbal Behavior, 6*(4), 238–252.

Ekman, P., Friesen, W. V., & O'Sullivan, M. (1988). Smiles when lying. *Journal of Personality and Social Psychology, 53*(3), 414–420.

Ekman, P., Friesen, W. V., & Scherer, K. R. (1976). Body movement and voice pitch in deceptive interaction. *Semiotica, 16*(1), 23–27.

Ekman, P., Levenson, R. W., & Friesen, W. V. (1983). Autonomic nervous system activity distinguishes among emotions. *Science, 221*(4616), 1208–1210.

Emojipedia. (n.d.-a). Face with tears of joy. Retrieved May 19, 2021 from https://emojipedia.org/face-with-tears-of-joy/

Emojipedia. (n.d.-b). Grinning face with big eyes. Retrieved May 10, 2021 from https://emojipedia.org/grinning-face-with-big-eyes/

Emojipedia. (n.d.-c). Smiling face with smiling eyes. Retrieved May 19, 2021 from https://emojipedia.org/smiling-face-with-smiling-eyes/

Epley, N., Waytz, A., & Cacioppo, J. T. (2007). On seeing human: A three-factor theory of anthropomorphism. *Psychological Review, 114*(4), 864–886.

Esteva, A., Kuprel, B., Novoa, R. A., Ko, J., Swetter, S. M., Blau, H. M., & Thrun, S. (2017). Dermatologist-level classification of skin cancer with deep neural networks. *Nature, 542*(February), 115–118.

Eyssel, F., & Hegel, F. (2012). (S)he's got the look: Gender stereotyping of robots. *Journal of Applied Social Psychology, 42*(9), 2213–2230.

Eyssel, F., Hegel, F., Horstmann, G., & Wagner, C. (2010). Anthropomorphic inferences from emotional nonverbal cues: A case study. In *Proceedings of the 19th International Symposium in Robot and Human Interactive Communication* (pp. 646–651). Viareggio, Italy: IEEE.

Featherman, M. S., Valacich, J. S., & Wells, J. D. (2006). Is that authentic or artificial? Understanding consumer perceptions of risk in e-service encounters. *Information Systems Journal, 16*(2), 107–134.

Felbo, B., Mislove, A., Søgaard, A., Rahwan, I., & Lehmann, S. (2017). Using millions of emoji occurrences to learn any-domain representations for detecting sentiment, emotion and sarcasm. In *Proceedings of the 2017 Conference on Empirical Methods in Natural Language Processing* (pp. 1615–1625). Copenhagen, Denmark: Association for Computational Linguistics.

Feshbach, N. D., & Roe, K. (1968). Empathy in six- and seven-year-olds. *Child Development, 39*(1), 133–145.

Fineman, S. (2004). Getting the measure of emotion-and the cautionary tale of emotional intelligence. *Human Relations, 57*(6), 719–740.

Fisher, C. D. (1997). *Emotions at work: What do people feel and how should we measure it?* (Discussion Paper No. 63). Gold Coast, Australia.

Forlizzi, J., Zimmerman, J., Mancuso, V., & Kwak, S. (2007). How interface agents affect interaction between humans and computers. In *Proceedings of the 2007 Conference on Designing Pleasurable Products and Interfaces* (pp. 209–221). Helsinki, Finland: Association for Computing Machinery.

Fornell, C., & Larcker, D. F. (1981). Evaluating structural equation models with unobservable variables and measurement error. *Journal of Marketing Research, 18*(1), 39–50.

Fowler, G. A. (2018). Want better customer service? Don't call. Text. Retrieved November 11, 2020 from https://www.washingtonpost.com/technology/2018/08/09/want-better-customer-service-dont-call-text/

Friedman, H. S., & Riggio, R. E. (1981). Effect of individual differences in nonverbal expressiveness on transmission of emotion. *Journal of Nonverbal Behavior, 6*(2), 96–104.

Frijda, N. H. (1986). *The emotions*. New York, USA: Cambridge University Press.

Gallagher, H. L., & Frith, C. D. (2003). Functional imaging of "theory of mind". *Trends in Cognitive Sciences, 7*(2), 77–83.

Gallese, V., Keysers, C., & Rizzolatti, G. (2004). A unifying view of the basis of social cognition. *Trends in Cognitive Sciences, 8*(9), 396–403.

Ganster, T., Eimler, S. C., & Krämer, N. C. (2012). Same same but different!? The differential influence of smilies and emoticons on person perception. *Cyberpsychology, Behavior, and Social Networking, 15*(4), 226–230.

Gardner, H. (1999). *Intelligence reframed: Multiple intelligences for the 21st century*. New York, USA: Basic Books.

Gefen, D., & Straub, D. W. (2004). Consumer trust in B2C e-Commerce and the importance of social presence: experiments in e-Products and e-Services. *Omega, 32*(6), 407–424.

Geisser, S. (1974). A predictive approach to the random effect model. *Biometrika, 61*(1), 101–107.

Genmod Project. (2016). Self-learning AI emulates the human brain (Sponsored by the European Union). Retrieved December 1, 2020 from https://ec.europa.eu/programmes/horizon2020/en/news/self-learning-ai-emulates-human-brain

Gentsch, P. (2018). *Künstliche Intelligenz für Sales, Marketing und Service [Artificial intelligence in sales, marketing, and service]*. Wiesbaden, Germany: Springer Fachmedien.

Geuens, M., & De Pelsmacker, P. (2017). Planning and conducting experimental advertising research and questionnaire design. *Journal of Advertising, 46*(1), 83–100.

Gill, A. J., Gergle, D., French, R. M., & Oberlander, J. (2008). Emotion rating from short blog texts. In *Proceedings of the SIGCHI Conference on Human Factors in Computing Systems* (pp. 1121–1124). Florence, Italy: Association for Computing Machinery.

Giuliani, M., Petrick, R., Foster, M. E., Gaschler, A., Isard, A., Pateraki, M., & Sigalas, M. (2013). Comparing task-based and socially intelligent behaviour in a robot bartender. In *Proceedings of the 15th ACM on International conference on multimodal interaction* (pp. 263–270). Sydney, Australia: Association for Computing Machinery.

Gladstein, G. A. (1983). Understanding empathy: Integrating counseling, developmental, and social psychology perspectives. *Journal of Counseling Psychology, 30*(4), 467–482.

Glomb, T. M., & Tews, M. J. (2004). Emotional labor: A conceptualization and scale development. *Journal of Vocational Behavior, 64*(1), 1–23.

Gnewuch, U., Adam, M. T. P., Morana, S., & Maedche, A. (2018). *The chatbot is typing ...'— The role of typing indicators in human-chatbot interaction*. Paper presented at the 17th Annual Pre-ICIS Workshop on HCI Research in MIS, San Francisco, USA.

Gnewuch, U., Morana, S., & Maedche, A. (2017). Towards designing cooperative and social conversational agents for customer service. In *Proceedings of the International Conference on Information Systems (ICIS) 2017* (pp. 1–13). Seoul, South Korea: Association for Information Systems.

Goldman, R., & Papson, S. (1996). *Sign wars the cluttered landscape of advertising*. New York, USA: Guildford.

Grandey, A. A. (2000). Emotional regulation in the workplace: A new way to conceptualize emotional labor. *Journal of Occupational Health Psychology, 5*(1), 95–110.

Grandey, A. A. (2003). When "the show must go on": Surface acting and deep acting as determinants of emotional exhaustion and peer-rated service delivery. *Academy of Management Journal, 46*(1), 86–96.

Grandey, A. A., Fisk, G. M., Mattila, A. S., Jansen, K. J., & Sideman, L. A. (2005). Is "service with a smile" enough? Authenticity of positive displays during service encounters. *Organizational Behavior and Human Decision Processes, 96*(1), 38–55.

Grégoire, Y., & Fisher, R. J. (2006). The effects of relationship quality on customer retaliation. *Marketing Letters, 17*(1), 31–46.

Gremler, D. D., & Gwinner, K. P. (2000). Customer-employee rapport in service relationships. *Journal of Service Research, 3*(1), 82–104.

Grewal, R., Cote, J. A., & Baumgartner, H. (2004). Multicollinearity and measurement error in structural equation models: Implications for theory testing. *Marketing Science, 23*(4), 519–529.

Griffin, B., & Hesketh, B. (2004). Why openness to experience is not a good predictor of job performance. *International Journal of Selection and Assessment, 12*(3), 243–251.

Gross, J. J. (1998). The emerging field of emotion regulation: An integrative review. *Review of General Psychology, 2*(3), 271–299.

Gross, J. J., Sutton, S. K., & Ketelaar, T. (1998). Relations between affect and personality: Support for the affect-level and affective-reactivity views. *Personality and Social Psychology Bulletin, 24*(3), 279–288.

Grossberg, S. (1969). Some networks that can learn, remember, and reproduce any number of complicated space-time patterns, I. *Journal of Mathematics and Mechanics, 19*(1), 53–91.

Groth, M., Hennig-Thurau, T., & Walsh, G. (2009). Customer reactions to emotional labor: The roles of employee acting strategies and customer detection accuracy. *Academy of Management Journal, 52*(5), 958–974.

Guthrie, S. E. (1993). *Faces in the clouds: A new theory of religion.* New York, USA: Oxford University Press.

Hair, J. F., Babin, B. J., Anderson, R. E., & Black, W. C. (2018a). *Multivariate data analysis* (8 ed.). Delhi, India: CENGAGE INDIA.

Hair, J. F., Hult, G. T. M., Ringle, C. M., & Sarstedt, M. (2014a). *A primer of partial least squares structural equation modeling (PLS-SEM).* Thousand Oaks, USA: Sage.

Hair, J. F., Risher, J. J., Sarstedt, M., & Ringle, C. M. (2019). When to use and how to report the results of PLS-SEM. *European Business Review, 31*(1), 2–24.

Hair, J. F., Sarstedt, M., Hopkins, L., & Kuppelwieser, V. G. (2014b). Partial least squares structural equation modeling (PLS-SEM): An emerging tool in business research. *European Business Review, 26*(2), 106–121.

Hair, J. F., Sarstedt, M., Ringle, C. M., & Gudergan, S. P. (2018b). *Advanced issues in partial least squares structural equation modeling (PLS-SEM).* Thousand Oaks, USA: Sage Publication.

Hair, J. F., Sarstedt, M., Ringle, C. M., & Mena, J. A. (2012). An assessment of the use of partial least squares structural equation modeling in marketing research. *Journal of the Academy of Marketing Science, 40*(3), 414–433.

Hall, L. (2005). Inflicting pain on synthetic characters: Moral concerns and empathic interaction. In *Proceedings of the Joint Symposium on Virtual Social Agents* (pp. 144–149). Hatfield, UK: The Society for the Study of Artificial Intelligence and the Simulation of Behaviour.

Hamid, O. H., Smith, N. L., & Barzanji, A. (2017). Automation, per se, is not job elimination: How artificial intelligence forwards cooperative human-machine coexistence. In *2017 IEEE 15th International Conference on Industrial Informatics (INDIN)* (pp. 899–904). Emden, Germany: IEEE.

Han, M. J., Lin, C. H., & Song, K. T. (2012). Robotic emotional expression generation based on mood transition and personality model. *IEEE Transactions on Cybernetics, 43*(4), 1290–1303.

Hancock, J. T., Gee, K., Ciaccio, K., & Lin, J. M.-H. (2008). I'm sad you're sad: Emotional contagion in CMC. In *Proceedings of the 2008 ACM Conference on Computer Supported Cooperative Work* (pp. 295–298). San Diego, USA: Association for Computing Machinery.

Hancock, J. T., Landrigan, C., & Silver, C. (2007). Expressing emotion in text-based communication. In *Proceedings of the SIGCHI Conference on Human Factors in Computing Systems* (pp. 929–932). San Jose, USA: Association for Computing Machinery.

Hatfield, E., Cacioppo, J. T., & Rapson, R. L. (1992). Primitive emotional contagion. In M. S. Clark (Ed.), *Emotion and social behavior* (pp. 151–177). Newbury Park, USA: SAGE Publications.

Hatfield, E., Cacioppo, J. T., & Rapson, R. L. (1993). Emotional contagion. *Current Directions in Psychological Science, 2*(3), 96–100.

Hatfield, E., Cacioppo, J. T., & Rapson, R. L. (1994). *Emotional contagion: Studies in emotion and social interaction.* Cambridge, UK: Cambridge University Press.

Hatfield, E., Carpenter, M., & Rapson, R. L. (2014). Emotional contagion as a precursor to collective emotions. In C. von Scheve & M. Salmela (Eds.), *Collective emotions* (pp. 108–122). Oxford, UK: Oxford University Press.

Hayes, A. F. (2017). *Introduction to mediation, moderation, and conditional process analysis: A regression-based approach* (2 ed.). New York, USA: Guilford Publications.

Heerink, M., Kröse, B., Evers, V., & Wielinga, B. (2010). Relating conversational expressiveness to social presence and acceptance of an assistive social robot. *Virtual Reality, 14*(1), 77–84.

Hegel, F., Krach, S., Kircher, T., Wrede, B., & Sagerer, G. (2008). Understanding social robots: A user study on anthropomorphism. In *RO-MAN 2008-The 17th IEEE International Symposium on Robot and Human Interactive Communication* (pp. 574–579). Munich, Germany: IEEE.

Heinen, N., Heuer, A., & Schautschick, P. (2017). Künstliche Intelligenz und der Faktor Arbeit [Artificial intelligence and the labor factor]. *Wirtschaftsdienst, 97*(10), 714–720.

Hennig-Thurau, T., Groth, M., Paul, M., & Gremler, D. D. (2006). Are all smiles created equal? How emotional contagion and emotional labor affect service relationships. *Journal of Marketing, 70*(3), 58–73.

Henseler, J., Ringle, C. M., & Sarstedt, M. (2015). A new criterion for assessing discriminant validity in variance-based structural equation modeling. *Journal of the Academy of Marketing Science, 43*(1), 115–135.

Henseler, J., Ringle, C. M., & Sarstedt, M. (2016). Testing measurement invariance of composites using partial least squares. *International Marketing Review, 33*(3), 405–431.

Henseler, J., Ringle, C. M., & Sinkovics, R. R. (2009). The use of partial least squares path modeling in international marketing. *Advances in International Marketing, 20*, 277–319.

Heskett, J. L., Sasser, W. E., & Schlesinger, L. A. (1997). *The service profit chain: How leading companies link profit and growth to loyalty, satisfaction, and value.* New York, USA: Free Press.

Hietanen, J. K., Surakka, V., & Linnankoski, I. (1998). Facial electromyographic responses to vocal affect expressions. *Psychophysiology, 35*(5), 530–536.

Higashinaka, R., Kobayashi, N., Hirano, T., Miyazaki, C., Meguro, T., Makino, T., & Matsuo, Y. (2016). Syntactic filtering and content-based retrieval of twitter sentences for the generation of system utterances in dialogue systems. In A. Rudnicky, A. Raux, I. Lane, & T.

Misu (Eds.), *Situated Dialog in Speech-Based Human-Computer Interaction* (pp. 15–26). Cham, Germany: Springer.
Hill, J., Ford, W. R., & Farreras, I. G. (2015). Real conversations with artificial intelligence: A comparison between human–human online conversations and human–chatbot conversations. *Computers in Human Behavior, 49*, 245–250.
Ho, A., Hancock, J., & Miner, A. S. (2018). Psychological, relational, and emotional effects of self-disclosure after conversations with a chatbot. *Journal of Communication, 68*(4), 712–733.
Hochschild, A. (1985). *The managed heart*. Berkeley, USA: University of California Press.
Hocutt, M. A., Bowers, M. R., & Donavan, D. T. (2006). The art of service recovery: fact or fiction? *Journal of Services Marketing, 20*(3), 199–207.
Holtgraves, T., & Han, T. L. (2007). A procedure for studying online conversational processing using a chat bot. *Behavior Research Methods, 39*(1), 156–163.
Homburg, N. (2018). How to include humanoid robots into experimental research: A multi-step approach. In *Proceedings of the 51st Hawaii International Conference on System Sciences* (pp. 4423–4432). Waikoloa Village, USA.
Horn, J. L., & McArdle, J. J. (1992). A practical and theoretical guide to measurement invariance in aging research. *Experimental Aging Research, 18*(3), 117–144.
Huang, A. H., Yen, D. C., & Zhang, X. (2008). Exploring the potential effects of emoticons. *Information & Management, 45*(7), 466–473.
Huang, M.-H., & Rust, R. T. (2018). Artificial intelligence in service. *Journal of Service Research, 21*(2), 155–172.
Huang, M.-H., Rust, R. T., & Maksimovic, V. (2019). The feeling economy: managing in the next generation of artificial intelligence (AI). *California Management Review, 61*(4), 43–65.
Hui, B. S., & Wold, H. (1982). Consistency and consistency at large of partial least squares estimates. In K. G. Jöreskog & H. Wold (Eds.), *Systems under indirect observation* (pp. 119–130). Amsterdam, The Netherlands: North Holland.
Hulland, J. (1999). Use of partial least squares (PLS) in strategic management research: a review of four recent studies. *Strategic Management Journal, 20*(2), 195–204.
Hulland, J., Baumgartner, H., & Smith, K. M. (2018). Marketing survey research best practices: evidence and recommendations from a review of JAMS articles. *Journal of the Academy of Marketing Science, 46*(1), 92–108.
Hulland, J., & Miller, J. (2018). „Keep on turkin". *Journal of the Academy of Marketing Science, 46*(5), 789–794.
Izard, C. E. (1977). *Human emotions*. New York, USA: Plenum Press.
Jackson, P. L., Brunet, E., Meltzoff, A. N., & Decety, J. (2006). Empathy examined through the neural mechanisms involved in imagining how I feel versus how you feel pain. *Neuropsychologia, 44*(5), 752–761.
Jain, M., Kumar, P., Kota, R., & Patel, S. N. (2018). Evaluating and informing the design of chatbots. In *Proceedings of the 2018 on Designing Interactive Systems Conference 2018* (pp. 895–906). Hong Kong, China: Association for Computing Machinery.
Jeffreys, D. A. (1993). The influence of stimulus orientation on the vertex positive scalp potential evoked by faces. *Experimental Brain Research, 96*(1), 163–172.

Jenkins, M. C., Churchill, R., Cox, S., & Smith, D. (2007). Analysis of user interaction with service oriented chatbot systems. In *Proceedings of the 12th International Conference on Human-Computer Interaction* (pp. 76–83). Beijing, China: Springer.

Johnson, H.-A. M., & Spector, P. E. (2007). Service with a smile: Do emotional intelligence, gender, and autonomy moderate the emotional labor process? *Journal of Occupational Health Psychology, 12*(4), 319–333.

Johnston, T. C., & Hewa, M. A. (1997). Fixing service failures. *Industrial Marketing Management, 26*(5), 467–473.

Jörling, M., Böhm, R., & Paluch, S. (2019). Service robots: Drivers of perceived responsibility for service outcomes. *Journal of Service Research, 22*(4), 404–420.

Jung, S., Lim, H. T., Kwak, S., & Biocca, F. (2012). Personality and facial expressions in human-robot interaction. In *Proceedings of the 7th ACM/IEEE International Conference on Human-Robot Interaction (HRI)* (pp. 161–162). Boston, USA: IEEE.

Kaplan, A., & Haenlein, M. (2019). Siri, Siri, in my hand: Who's the fairest in the land? On the interpretations, illustrations, and implications of artificial intelligence. *Business Horizons, 62*(1), 15–25.

Karmarkar, U. (2004). Will you survive the services revolution? *Harvard Business Review, 82*(6), 100–107.

Katagiri, Y., Nass, C., & Takeuchi, Y. (2001). Cross-cultural studies of the computers are social actors paradigm: The case of reciprocity. In M. J. Smith, G. Salvendy, D. Harris, & R. Koubek (Eds.), *Usability evaluation and interface design: Cognitive engineering, intelligent agents, and virtual reality* (pp. 1558–1562). Mahwah, USA: Lawrence Erlbaum.

Keh, H. T., Ren, R., Hill, S. R., & Li, X. (2013). The beautiful, the cheerful, and the helpful: The effects of service employee attributes on customer satisfaction. *Psychology & Marketing, 30*(3), 211–226.

Keltner, D., & Gross, J. J. (1999). Functional accounts of emotions. *Cognition & Emotion, 13*(5), 467–480.

Kenning, P., Plassmann, H., Deppe, M., Kugel, H., & Schwindt, W. (2005). Wie eine starke Marke wirkt [How strong brands impact]. *Harvard Business Manager*(March), 53–57.

Keysers, C., & Gazzola, V. (2006). Towards a unifying neural theory of social cognition. *Progress in Brain Research, 156*, 379–401.

Kiesler, S., Siegel, J., & McGuire, T. W. (1984). Social psychological aspects of computer-mediated communication. *American Psychologist, 39*(10), 1123–1134.

Kim, J., & Moon, J. Y. (1998). Designing towards emotional usability in customer interfaces—trustworthiness of cyber-banking system interfaces. *Interacting with Computers, 10*(1), 1–29.

Kim, K. J., Park, E., & Sundar, S. S. (2013). Caregiving role in human–robot interaction: A study of the mediating effects of perceived benefit and social presence. *Computers in Human Behavior, 29*(4), 1799–1806.

Klein, J., Moon, Y., & Picard, R. W. (2002). This computer responds to user frustration: Theory, design, and results. *Interacting with Computers, 14*(2), 119–140.

Kline, R. B. (2011). *Principles and practice of structural equation modeling*. New York, USA: Guilford Press.

Koch, M. (2018). Artificial intelligence is becoming natural. *Cell, 173*(3), 531–533.

Kopp, S., Gesellensetter, L., Krämer, N. C., & Wachsmuth, I. (2005). A conversational agent as museum guide—Design and evaluation of a real-world application. In T.

Panayiotopoulos, J. Gratch, R. Aylett, D. Ballin, P. Olivier, & T. Rist (Eds.), *Intelligent virtual agents. IVA 2005. Lecture notes in computer science* (Vol. 3661, pp. 329–343). Berlin, Germany: Springer.

Kuchenbrandt, D., Häring, M., Eichberg, J., Eyssel, F., & André, E. (2014). Keep an eye on the task! How gender typicality of tasks influence human–robot interactions. *International Journal of Social Robotics, 6*(3), 417–427.

Kwak, S. S., Kim, Y., Kim, E., Shin, C., & Cho, K. (2013). What makes people empathize with an emotional robot?: The impact of agency and physical embodiment on human empathy for a robot. In *2013 IEEE RO-MAN* (pp. 180–185). Gyeongju, South Korea: IEEE.

Kwidzinski, R. (2020). So positioniert sich Roboter Jeeves in Corona-Zeiten [This is how robot Jeeves positions itself in Corona times]. Retrieved January 6, 2021 from https://www.ahgz.de/hotellerie/news/so-positioniert-sich-roboter-jeeves-in-corona-zeiten-275967

La, S., & Choi, B. (2019). Perceived justice and CSR after service recovery. . *Journal of Services Marketing, 33*(2), 206–219.

Lang, F. R., John, D., Lüdtke, O., Schupp, J., & Wagner, G. G. (2011). Short assessment of the Big Five: Robust across survey methods except telephone interviewing. *Behavior Research Methods, 43*(2), 548–567.

Larivière, B., Bowen, D., Andreassen, T. W., Kunz, W., Sirianni, N. J., Voss, C., . . . De Keyser, A. (2017). "Service Encounter 2.0": An investigation into the roles of technology, employees and customers. *Journal of Business Research, 79*, 238–246.

Larsen, R. J., & Ketelaar, T. (1989). Extraversion, neuroticism and susceptibility to positive and negative mood induction procedures. *Personality and Individual Differences, 10*(12), 1221–1228.

Latah, M., & Toker, L. (2018). Artificial intelligence enabled software-defined networking: a comprehensive overview. *IET Networks, 8*(2), 79–99.

Lazarus, R. S. (1991). Progress on a cognitive-motivational-relational theory of emotion. *American Psychologist, 46*(8), 819–834.

Lebovits, G. (2015). E-mail netiquette for lawyers. *New York State Bar Association Journal, 81*(9), 56–59.

LeCun, Y., Bengio, Y., & Hinton, G. (2015). Deep learning. *Nature, 521*(7553), 436–444.

Lee, K. M. (2004). Presence, explicated. *Communication Theory, 14*(1), 27–50.

Lee, K. M., Peng, W., Jin, S. A., & Yan, C. (2006). Can robots manifest personality?: An empirical test of personality recognition, social responses, and social presence in human–robot interaction. *Journal of Communication, 56*(4), 754–772.

Lee, M. J., Singh, N., & Chan, E. S. (2011). Service failures and recovery actions in the hotel industry: A text-mining approach. *Journal of Vacation Marketing, 17*(3), 197–207.

Lee, M. K., Kiesler, S., & Forlizzi, J. (2010). Receptionist or information kiosk: How do people talk with a robot? In *Proceedings of the 2010 ACM Conference on Computer Supported Cooperative Work* (pp. 31–40). Savannah, USA: Association for Computing Machinery.

Levenson, R. W., Ekman, P., & Friesen, W. V. (1990). Voluntary facial action generates emotion-specific autonomic nervous system activity. *Psychophysiology, 27*(4), 363–384.

Lewinski, P., den Uyl, T. M., & Butler, C. (2014). Automated facial coding: validation of basic emotions and FACS AUs in FaceReader. *Journal of Neuroscience, Psychology, and Economics, 7*(4), 227–236.

Li, X., Chan, K. W., & Kim, S. (2019). Service with emoticons: How customers interpret employee use of emoticons in online service encounters. *Journal of Consumer Research, 45*(5), 973–987.

Li, Y., Su, H., Shen, X., Li, W., Cao, Z., & Niu, S. (2017). *Dailydialog: A manually labelled multi-turn dialogue dataset.* Paper presented at the International Joint Conference on Natural Language Processing, Taipei, Taiwan.

Liao, Q. V., Davis, M., Geyer, W., Muller, M., & Shami, N. S. (2016). What can you do?: Studying social-agent orientation and agent proactive interactions with an agent for employees. In *Proceedings of the 2016 ACM Conference on Designing Interactive Systems (DIS '16)* (pp. 264–275). New York, USA: Association for Computing Machinery.

Liao, Q. V., Hussain, M. M., Chandar, P., Davis, M., Crasso, M., Wang, D., . . . Geyer, W. (2018). *All work and no play? Conversations with a question-and-answer chatbot in the wild.* Paper presented at the 2018 CHI Conference on Human Factors in Computing Systems, Montreal, Canada.

Lieberman, M. D. (2000). Intuition: A social cognitive neuroscience approach. *Psychological Bulletin, 126*(1), 109–137.

Lin, Y.-T., Doong, H.-S., & Eisingerich, A. B. (2021). Avatar design of virtual salespeople: Mitigation of recommendation conflicts. *Journal of Service Research, 24*(1), 141–159.

Lindell, M. K., & Whitney, D. J. (2001). Accounting for common method variance in cross-sectional research designs. *Journal of Applied Psychology, 86*(1), 114–121.

Lipps, T. (1907). Das Wissen von fremden Ichen [The knowledge of others I]. *Psychologische Untersuchungen, 1*(4), 694–722.

Liu, B., & Sundar, S. S. (2018). Should machines express sympathy and empathy? Experiments with a health advice chatbot. *Cyberpsychology, Behavior, and Social Networking, 21*(10), 625–636.

Lohmann, K., Pyka, S. S., & Zanger, C. (2017). The effects of smileys on receivers' emotions. *Journal of Consumer Marketing, 34*(6), 489–495.

Lohmoeller, J. B. (1989). *Latent variable path modeling with partial least squares.* Heidelberg, Germany: Physica.

Luger, E., & Sellen, A. (2016). Like having a really bad PA: the gulf between user expectation and experience of conversational agents. In *Proceedings of the 2016 CHI Conference on Human Factors in Computing Systems* (pp. 5286–5297). San Jose, USA: Association for Computing Machinery.

Lukanova, G., & Ilieva, G. (2019). Robots, artificial intelligence and service automation in hotels. In S. Ivanov & C. Webster (Eds.), *Robots, Artificial Intelligence, and Service Automation in Travel, Tourism, and Hospitality* (pp. 157–183). Bingley, UK: Emerald Publishing Limited.

Luo, X., Tong, S., Fang, Z., & Qu, Z. (2019). Frontiers: Machines vs. humans: The impact of artificial intelligence chatbot disclosure on customer purchases. *Marketing Science, 38*(6), 937–947.

Mackiewicz, J. (2003). Which rules for online writing are worth following?: A study of eight rules in eleven handbooks. *IEEE Transactions on Professional Communication, 46*(2), 129–137.

Marcoulides, G. A., Chin, W. W., & Saunders, C. (2012). When imprecise statistical statements become problematic: a response to Goodhue, Lewis, and Thompson. *MIS Quarterly, 36*(3), 717–728.

Marinova, D., de Ruyter, K., Huang, M.-H., Meuter, M. L., & Challagalla, G. (2017). Getting smart: Learning from technology-empowered frontline interactions. *Journal of Service Research, 20*(1), 29–42.

Martin, D., O'Neill, M., Hubbard, S., & Palmer, A. (2008). The role of emotion in explaining consumer satisfaction and future behavioural intention. *Journal of Services Marketing, 22*(3), 224–236.

Matthews, G. (1997). Extraversion, emotion and performance: A cognitive-adaptive model. *Advances in Psychology, 124*, 399–442.

Matthews, L. (2017). Applying multigroup analysis in PLS-SEM: A step-by-step process. In H. Latan & R. Noonan (Eds.), *Partial least squares path modeling* (pp. 219–243). Cham, Germany: Springer.

Mattingly, V., & Kraiger, K. (2019). Can emotional intelligence be trained? A meta-analytical investigation. *Human Resource Management Review, 29*(2), 140–155.

Maxham III, J. G., & Netemeyer, R. G. (2002). Modeling customer perceptions of complaint handling over time: the effects of perceived justice on satisfaction and intent. *Journal of Retailing, 78*(4), 239–252.

McCarthy, J. (2007). From here to human-level AI. *Artificial Intelligence, 171*(18), 1174–1182.

McCarthy, J., Minsky, M. L., Rochester, N., & Shannon, C. E. (2006). A proposal for the dartmouth summer research project on artificial intelligence (August 31, 1955). *AI Magazine, 27*(4), 12–14.

McClure, S. M., Li, J., Tomlin, D., Cypert, K. S., Montague, L. M., & Montague, P. R. (2004). Neural correlates of behavioral preference for culturally familiar drinks. *Neuron, 44*, 379–387.

McCollough, M. A., & Sundar, G. B. (1992). The recovery paradox: An examination of customer satisfaction in relation to disconfirmation, service quality, and attribution based theories. In C. T. Allen (Ed.), *Marketing theory and applications*. Chicago, USA: American Marketing Association.

McCrae, R. R., & Costa, P. T. (1997). Conceptions and correlates of openness to experience. In R. Hogan, J. Johnson, & S. Briggs (Eds.), *Handbook of personality psychology* (pp. 825–847). Cambridge: Academic Press.

McCrae, R. R., & Terracciano, A. (2005). Universal features of personality traits from the observer's perspective: data from 50 cultures. *Journal of Personality and Social Psychology, 88*(3), 547–561.

McCulloch, W. S., & Pitts, W. (1943). A logical calculus of the ideas immanent in nervous activity. *The Bulletin of Mathematical Biophysics, 5*(4), 115–133.

McHugo, G. J., Lanzetta, J. T., Sullivan, D. G., Masters, R. D., & Englis, B. G. (1985). Emotional reactions to a political leader's expressive displays. *Journal of Personality and Social Psychology, 49*(6), 1513–1529.

McNiel, J. M., & Fleeson, W. (2006). The causal effects of extraversion on positive affect and neuroticism on negative affect: Manipulating state extraversion and state neuroticism in an experimental approach. *Journal of Research in Personality, 40*(5), 529–550.

McTear, M., Callejas, Z., & Griol, D. (2016). *The conversational interface: Talking to smart devices*. Basel, Switzerland: Springer International Publishing.

Mead, G. H. (1934). *Mind, self and society from the standpoint of a social behaviorist*. Chicago, USA: University of Chicago Press.

Medler-Liraz, H. (2016). The role of service relationships in employees' and customers' emotional behavior, and customer-related outcomes. *Journal of Services Marketing, 30*(4), 437–448.

Mehrabian, A., & Epstein, N. (1972). A measure of emotional empathy. *Journal of Personality, 40*(4), 525–543.

Mesmer-Magnus, J. R., DeChurch, L. A., & Wax, A. (2012). Moving emotional labor beyond surface and deep acting. *Organizational Psychology Review, 2*(1), 6–53.

Meuter, M. L., & Bitner, M. J. (1998). Self-service technologies: Extending service frameworks and identifying issues for research. In D. Grewal & C. Pechman (Eds.), *Marketing theory and applications* (Vol. 9, pp. 12–19). Chicago, USA: American Marketing Association.

Miller, H., Thebault-Spieker, J., Chang, S., Johnson, I., Terveen, L., & Hecht, B. (2016). *„Blissfully happy" or „ready to fight": Varying interpretations of emoji.* Paper presented at the International AAAI Conference on Web and Social Media, Cologne, Germany.

Minsky, M. (2006). *The emotion machine.* New York, USA: Simon & Schuster.

Mongin, X. (2020). Technology is vital to the hospitality industry revival. Retrieved March 19, 2021 from https://www.hospitalitynet.org/opinion/4099077.html

Moon, Y. (1999). The effects of physical distance and response latency on persuasion in computer-mediated communication and human-computer communication. *Journal of Experimental Psychology: Applied, 5*(4), 379–392.

Moor, P. J., Heuvelman, A., & Verleur, R. (2010). Flaming on youtube. *Computers in Human Behavior, 26*(6), 1536–1546.

Morrissey, K., & Kirakowski, J. (2013). 'Realness' in chatbots: Establishing quantifiable criteria. In *Proceedings of the 15th International Conference on Human-Computer Interaction* (pp. 87–96). Las Vegas, USA: Springer.

Mowrer, O. H. (1960). *Learning theory and behavior*. New York, USA: Wiley.

Mozafari, N., Weiger, W. H., & Hammerschmidt, M. (2020). The chatbot disclosure dilemma: Desirable and undesirable effects of disclosing the non-human identity of chatbots. In *Proceedings of the 41st International Conference on Information Systems.*

Murgia, A., Janssens, D., Demeyer, S., & Vasilescu, B. (2016). Among the machines: Human-bot interaction on social Q&A websites. In *Proceedings of the 2016 CHI Conference Extended Abstracts on Human Factors in Computing Systems* (pp. 1272–1279). San Jose, USA: Association for Computing Machinery.

Mustak, M., Salminen, J., Plé, L., & Wirtz, J. (2021). Artificial intelligence in marketing: Topic modeling, scientometric analysis, and research agenda. *Journal of Business Research, 124*, 389–404.

Myers, D. G. (1989). *Psychology* (2nd ed.). New York: Worth Publishers.

Nass, C., & Moon, Y. (2000). Machines and mindlessness: Social responses to computers. *Journal of Social Issues, 56*(1), 81–103.

Neumann, R., & Strack, F. (2000). „Mood contagion": The automatic transfer of mood between persons. *Journal of Personality and Social Psychology, 79*(2), 211–223.

Nilsson, N. J. (2005). Human-level artificial intelligence? Be serious! *AI Magazine, 26*(4), 68–75.

Nitzl, C., Roldán, J. L., & Cepeda Carrión, G. (2016). Mediation analysis in partial least squares path modeling: Helping researchers discuss more sophisticated models. *Industrial Management & Data Systems, 119*(9), 1849–1864.

Nunamaker, J. F., Derrick, D. C., Elkins, A. C., Burgoon, J. K., & Patton, M. W. (2011). Embodied conversational agent-based kiosk for automated interviewing. *Journal of Management Information Systems, 28*(1), 17–48.

Ochsner, K., & Gross, J. (2005). The cognitive control of emotion. *Trends in Cognitive Sciences, 9*(5), 242–249.

Oliver, R. L. (1981). Measurement and evaluation of satisfaction process in retail setting. *Journal of Retailing, 57*, 25–48.

Oliver, R. L. (1993). Cognitive, affective, and attribute bases of the satisfaction response. *Journal of Consumer Research, 20*(3), 418–430.

Olszanowski, M., Pochwatko, G., Kukliński, K., Ścibor-Rylski, M., & Ohme, R. (2008). *Warsaw set of emotional facial expression pictures—Validation study*. Paper presented at the EAESP General Meeting, Opatija, Croatia.

Oracle. (2016). Can virtual experiences replace reality? Retrieved May 20, 2019 from https://www.oracle.com/webfolder/s/delivery_production/docs/FY16h1/doc35/CXResearchVirtualExperiences.pdf

Oxford Dictionary. Definition emoji. Retrieved October 28, 2020 from https://www.oxfordlearnersdictionaries.com/definition/english/emoji?q=emoji

Paiva, A., Dias, J., Sobral, D., Woods, S., & Hall, L. (2004). *Building empathic lifelike characters: The proximity factor.* Paper presented at the International Joint Conference on Autonomous Agents and Multi-Agent Systems (AAMAS), New York.

Parasuraman, A., Zeithaml, V. A., & Berry, L. L. (1988). SERVQUAL: A multiple-item scale for measuring consumer perceptions of service quality. *Journal of Retailing, 64*(1), 12–40.

Pascual-Leone, A., Herpertz, S. C., & Kramer, U. (2016). Experimental designs and the ‚emotion stimulus critique': Hidden problems and potential solutions in the study of emotion. *Psychopathology, 49*(1), 60–68.

Patterson, P. G., McColl-Kennedy, J. R., Smith, A. K., & Lu, Z. (2009). Customer rage: triggers, tipping points, and take-outs. *California Management Review, 52*(1), 6–28.

Perdue, B. C., & Summers, J. O. (1986). Checking the success of manipulations in marketing experiments. *Journal of Marketing Research, 23*(4), 317–326.

Phillips, M. L., Drevets, W. C., Rauch, S. L., & Lane, R. (2003). Neurobiology of emotion perception I: The neural basis of normal emotion perception. *Biological Psychiatry, 54*(5), 504–514.

Picard, R. W. (1995, 1995). *Affective computing*, Cambridge, USA.

Plutchik, R. (1962). *The emotions: Facts, theories, and a new model.* New York, USA: Random House.

Podsakoff, P. M., MacKenzie, S. B., Lee, J.-Y., & Podsakoff, N. P. (2003). Common method biases in behavioral research: A critical review of the literature and recommended remedies. *Journal of Applied Psychology, 88*(5), 879–903.

Podsakoff, P. M., MacKenzie, S. B., & Podsakoff, N. P. (2012). Sources of method bias in social science research and recommendations on how to control it. *Annual Review of Psychology, 63*, 539–569.

Preston, S. D., & De Waal, F. B. (2002). Empathy: Its ultimate and proximate bases. *Behavioral and Brain Sciences, 25*(1), 1–20.

Prochazkova, E., & Kret, M. E. (2017). Connecting minds and sharing emotions through mimicry: A neurocognitive model of emotional contagion. *Neuroscience & Biobehavioral Reviews, 80*, 99–114.

Pugh, S. D. (2001). Service with a smile: Emotional contagion in the service encounter. *Academy of Management Journal, 44*(5), 1018–1027.

Rafaeli, A., & Sutton, R. I. (1987). Expression of emotion as part of the work role. *Academy of Management Review, 12*(1), 23–37.

Rammstedt, B., & John, O. P. (2005). Kurzversion des big five inventory (BFI-K). *Diagnostica, 51*(4), 195–206.

Ramos, C., Augusto, J. C., & Shapiro, D. (2008). Ambient intelligence—the next step for artificial intelligence. *IEEE Intelligent Systems, 23*(2), 15–18.

Reeves, B., & Nass, C. I. (1996). *The Media Equation: How people treat computers, television, and new media like real people and places*. Cambridge: Cambridge University Press.

Reinartz, W. J., Haenlein, M., & Henseler, J. (2009). An empirical comparison of the efficacy of covariance-based and variance-based SEM. *International Journal of Market Research, 26*(4), 332–344.

Rezabek, L., & Cochenour, J. (1998). Visual cues in computer-mediated communication: supplementing text with emoticons. *Journal of Visual Literacy, 18*(2), 201–215.

Riek, L. D., Rabinowitch, T. C., Chakrabarti, B., & Robinson, P. (2009a). Empathizing with robots: Fellow feeling along the anthropomorphic spectrum. In *2009 3rd International Conference on Affective Computing and Intelligent Interaction and Workshops* (pp. 1–6). Amsterdam, The Netherlands: IEEE.

Riek, L. D., Rabinowitch, T. C., Chakrabarti, B., & Robinson, P. (2009b). How anthropomorphism affects empathy toward robots. In *Proceedings of the 4th ACM/IEEE International Conference on Human Robot Interaction* (pp. 245–246). La Jolla, USA: Association for Computing Machinery.

Riikkinen, M., Saarijärvi, H., Sarlin, P., & Lähteenmäki, I. (2018). Using artificial intelligence to create value in insurance. *International Journal of Bank Marketing, 36*(6), 1145–1168.

Ringle, C. M., Wende, S., & Becker, J.-M. (2015). *SmartPLS 3*. Boenningstedt: SmartPLS GmbH.

Rizzolatti, G., Fogassi, L., & Gallese, V. (2001). Neurophysiological mechanisms underlying the understanding and imitation of action. *Nature Reviews Neuroscience, 2*(9), 661–670.

Roeyers, H., Buysse, A., Ponnet, K., & Pichal, B. (2001). Advancing advanced mind-reading tests: empathic accuracy in adults with a pervasive developmental disorder. *Journal of Child Psychology and Psychiatry, and Allied Disciplines, 42*(2), 271–278.

Rogers, C. (1961). *On becoming a person: A therapist's view of psychotherapy*. Boston, USA: Houghton Mifflin.

Rogers, C. R. (1959). A theory of therpy, personality and interpersonal relationships as developed in the client-centered framework. In S. Koch (Ed.), *Psychology: a study of a science, vol. III. Formulations of the person and the social context* (pp. 184–256). New York, USA: McGraw Hill.

Rogers, K., Dziobek, I., Hassenstab, J., Wolf, O. T., & Convit, A. (2007). Who cares? Revisiting empathy in Asperger syndrome. *Journal of Autism and Developmental Disorders, 37*(4), 709–715.

Rosenblatt, F. (1985). The perceptron: a probabilistic model for information storage and organization in the brain. *Psychological Review, 65*(6), 386–408.

Rosenthal-von der Pütten, A. M., Krämer, N. C., Hoffmann, L., Sobieraj, S., & Eimler, S. C. (2013). An experimental study on emotional reactions towards a robot. *International Journal of Social Robotics, 5*(1), 17–34.

Rosenthal-von der Pütten, A. M., Schulte, F. P., Eimler, S. C., Sobieraj, S., Hoffmann, L., Maderwald, S., . . . Krämer, N. C. (2014). Investigations on empathy towards humans and robots using fMRI. *Computers in Human Behavior, 33*(April), 201–212.

Ruby, P., & Decety, J. (2004). How would you feel versus how do you think she would feel? A neuroimaging study of perspective-taking with social emotions. *Journal of Cognitive Neuroscience, 16*(6), 988–999.

Ruff, C. C., & Fehr, E. (2014). The neurobiology of rewards and values in social decision making. *Nature Reviews Neuroscience, 15*(8), 549–562.

Rüggenberg, S. (2007). So nah und doch so fern—Soziale Präsenz und Vertrauen in der computervermittelten Kommunikation [So near and yet so far—Social presence and trust in computer-mediated communication]. *Dissertation, Department of Psychology, University of Cologne, Cologne, Germany.*

Russell, J. A. (1980). A circumplex model of affect. *Journal of Personality and Social Psychology, 39*(6), 1161–1178.

Russell, S. J., & Norvig, P. (2010). *Artificial intelligence: A modern approach* (3rd ed.). Essex, UK: Pearson.

Salovey, P., & Mayer, J. D. (1990). Emotional intelligence. *Imagination, Cognition and Personality, 9*(3), 185–211.

Sarikaya, R. (2017). The technology behind personal digital assistants: An overview of the system architecture and key components. *IEEE Signal Processing Magazine, 34*(1), 67–81.

Sasaki, M. U., & Ohbuchi, K. (1999). Conflict processes on computer-mediated communication. *Tohoku Psychologica Folia, 58*, 50–55.

Sawhney, M. (2016). Putting products into services. *Harvard Business Review*(September), 82–89.

Sawyer, A. G., & Ball, A. D. (1981). Statistical power and effect size in marketing research. *Journal of Marketing Research, 18*(3), 275–290.

Schanke, S., Burtch, G., & Ray, G. (2021). Estimating the impact of 'humanizing' customer service chatbots. *Information Systems Research, 32*(3), 736–751.

Schaufeli, W. B., Bakker, A. B., & Rhenen, W. V. (2009). How changes in job demands and resources predict burnout, work engagement, and sickness absenteeism. *Journal of Organizational Behavior, 30*(7), 893–917.

Scheflen, A. E. (1964). The significance of posture in communication systems. *Psychiatry, 27*(4), 316–331.

Scheler, M. (1948). *Wesen und Formen der Sympathie [Nature and forms of sympathy]* (Vol. 5). Frankfurt a. M., Germany: Schulte-Bulmke.

Schlobinski, P., & Siever, T. (2018). *Sprachliche Kommunikation in der digitalen Welt [Linguistic communication in the digital world]*. Berlin, Germany: forsa Politik- und Sozialforschung GmbH, Gesellschaft für deutsche Sprache e. V., mediensprache.net.

Schmidhuber, J. (2015). Deep learning in neural networks: An overview. *Neural Networks, 61*, 85–117.

Schneider, C. (2017). 10 Reasons why AI-powered, automated customer service is the future. Retrieved November 27, 2018 from https://www.ibm.com/blogs/watson/2017/10/10-reasons-ai-powered-automated-customer-service-future/

Schneider, R., Leibinger, C., Jentzsch, A., Leite, I., Paiva, A., Zoll, C., & Enz, S. (2012). Does empathic robot behavior enhance positive emotions in child-robot interaction? In *Proceedings of the 3rd International Workshop Teaching Robotics, Teaching with Robotics: Integrating Robotics* (pp. 200–201). Trento, Italy.

Schutte, N. S., Malouff, J. M., Bobik, C., Coston, T. D., Greeson, C., Jedlicka, C., . . . Wendorf, G. (2001). Emotional intelligence and interpersonal relations. *The Journal of Social Psychology, 141*(4), 523–536.

Schwarz, N., & Clore, G. L. (1983). Mood, misattribution, and judgments of well-being: informative and directive functions of affective states. *Journal of Personality and Social Psychology, 45*(3), 513–523.

Searle, J. R. (1980). Minds, brains and programs. *Behavioral and Brain Sciences, 3*, 417–457.

Sebes, C. (2018). How artificial intelligence is reshaping the restaurant world. Retrieved November 17, 2020 from https://www.qsrmagazine.com/outside-insights/how-artificial-intelligence-reshaping-restaurant-world

Serban, I. V., Klinger, T., Tesauro, G., Talamadupula, K., Zhou, B., Bengio, Y., & Courville, A. (2017). Multiresolution recurrent neural networks: An application to dialogue response generation. In *Proceedings of the Thirty-First AAAI Conference on Artificial Intelligence* (pp. 3288–3294). San Francisco, USA: AAAI Press.

Shechtman, N., & Horowitz, L. M. (2003). Media inequality in conversation: How people behave differently when interacting with computers and people. In *Proceedings of the SIGCHI Conference on Human Factors in Computing Systems* (pp. 281–288). Florida, USA: Association for Computing Machinery.

Shi, W., Wang, X., Oh, Y. J., Zhang, J., Sahay, S., & Yu, Z. (2020). Effects of persuasive dialogues: Testing bot identities and inquiry strategies. In *Proceedings of the 2020 CHI Conference on Human Factors in Computing Systems* (pp. 1–13). Honolulu, USA: Association for Computing Machinery.

Shiota, M. N., Keltner, D., & John, O. P. (2006). Positive emotion dispositions differentially associated with Big Five personality and attachment style. *The Journal of Positive Psychology, 1*(2), 61–71.

Shiv, B., & Fedorikhin, A. (1999). Heart and mind in conflict: The interplay of affect and cognition in consumer decision making. *Journal of Consumer Research, 26*(3), 278–292.

Shum, H.-Y., He, X.-D., & Li, D. (2018). From Eliza to XiaoIce: Challenges and opportunities with social chatbots. *Frontiers of Information Technology & Electronic Engineering, 19*(1), 10–26.

Siegel, J., Dubrovsky, V., Kiesler, S., & McGuire, T. W. (1986). Group processes in computer-mediated communication. *Organizational Behavior and Human Decision Processes, 37*(2), 157–187.

Sigall, H., & Mills, J. (1998). Measures of independent variables and mediators are useful in social psychology experiments: But are they necessary? *Personality and Social Psychology Review, 2*(3), 218–226.

Simmering, M. J., Fuller, C. M., Richardson, H. A., Ocal, Y., & Atinc, G. M. (2015). Marker variable choice, reporting, and interpretation in the detection of common method variance: A review and demonstration. *Organizational Research Methods, 18*(3), 473–511.

Singh, J. (1990). Voice, exit, and negative word-of-mouth behaviors: An investigation across three service categories. *Journal of the Academy of Marketing Science, 18*(1), 1–15.

Sion, G. (2018). How artificial intelligence is transforming the economy. Will cognitively enhanced machines decrease and eliminate tasks from human workers through automation? *Journal of Self-Governance and Management Economics, 6*(4), 31–36.

Skjuve, M., Haugstveit, I. M., Følstad, A., & Brandtzaeg, P. B. (2019). Help! Is my chatbot falling into the Uncanny Valley? An empirical study of user experience in human-chatbot interaction. *Human Technology, 15*(1), 30–54.

Smith, A. K., & Bolton, R. N. (2002). The effect of customers' emotional responses to service failures on their recovery effort evaluations and satisfaction judgments. *Journal of the Academy of Marketing Science, 30*(1), 5–23.

Smith, A. K., Bolton, R. N., & Wagner, J. (1999). A model of customer satisfaction with service encounters involving failure and recovery. *Journal of Marketing Research, 36*(3), 356–372.

Smith, L. W., & Rose, R. L. (2020). Service with a smiley face: Emojional contagion in digitally mediated relationships. *International Journal of Research in Marketing, 37*(2), 301–319.

Spencer, G. (2018). Much more than a chatbot: China's Xiaoice mixes AI with emotions and wins over millions of fans. Retrieved December 9, 2020 from https://news.microsoft.com/apac/features/much-more-than-a-chatbot-chinas-xiaoice-mixes-ai-with-emotions-and-wins-over-millions-of-fans/

Squires, A., Wade, J., Dominick, P., & Gelosh, D. (2011). Building a competency taxonomy to guide experience acceleration of lead program systems engineers. In *9th Annual Conference on Systems Engineering Research (CSER)* (pp. 1–10). Redondo Beach, USA.

Steger, J. (2017). Im Zwiegespräch mit den Robotern [In dialogue with the robots]. Retrieved October 27, 2020 from https://www.handelsblatt.com/unternehmen/it-medien/unternehmen-setzen-auf-chatbots-im-zwiegespraech-mit-den-robotern/19968428.html

Sternberg, R. J. (1999). The theory of successful intelligence. *Review of General Psychology, 3*(4), 292–316.

Sternberg, R. J. (2005). The theory of successful intelligence. *Interamerican Journal of Psychology, 39*(2), 189–202.

Stock, R. M. (2018). Can service robots hamper customer anger and aggression after a service failure? *Department of Business Administration, Economics and Law, Institute for Business Studies (BWL), Darmstadt Technical University, Darmstadt, Germany.*

Stock, R. M., & Merkle, M. (2018). Can humanoid service robots perform better than service employees? A comparison of innovative behavior cues. In *Proceedings of the 51st Hawaii International Conference on System Sciences* (pp. 1056–1065). Waikoloa Village, USA.

Stöckli, S., Schulte-Mecklenbeck, M., Borer, S., & Samson, A. C. (2018). Facial expression analysis with AFFDEX and FACET: A validation study. *Behavior Research Methods, 50*(4), 1446–1460.

Stoll, M., Baecke, S., & Kenning, P. (2008). What they see is what they get? An fMRI-study on neural correlates of attractive packaging. *Journal of Consumer Behaviour, 7*(4), 342–359.

Stone, M. (1974). Cross-validatory choice and assessment of statistical predictions. *Journal of the Royal Statistical Society, 36*(2), 111–147.

Sullins, E. S. (1991). Emotional contagion revisited: Effects of social comparison and expressive style on mood convergence. *Personality and Social Psychology Bulletin, 17*(2), 166–174.

Tellegen, A. (1985). Structures of mood and personality and their relevance to assessing anxiety, with an emphasis on self-report. In A. H. Tuma & J. D. Maser (Eds.), *Anxiety and the anxiety disorders* (pp. 681–706). Hillsdale, USA: Lawrence Erlbaum Associates.

Thies, I. M., Menon, N., Magapu, S., Subramony, M., & O'Neill, J. (2017). How do you want your chatbot? An exploratory Wizard-of-Oz study with young, urban Indians. In *Proceedings of the 16th IFIP TC 13 International Conference* (pp. 441–459). Mumbai, India: Springer.

Tielman, M., Neerincx, M., Meyer, J.-J., & Looije, R. (2014, 2014). *Adaptive emotional expression in robot-child interaction*, New York, USA.

Tsai, J., Bowring, E., Marsella, S., Wood, W., & Tambe, M. (2012). *A Study of emotional contagion with virtual characters*, Berlin, Germany.

Tsai, W.-C., & Huang, Y.-M. (2002). Mechanisms linking employee affective delivery and customer behavioral intentions. *Journal of Applied Psychology, 87*(5), 1001–1008.

Tsaur, S. H., & Lin, Y. C. (2004). Promoting service quality in tourist hotels: the role of HRM practices and service behavior. *Tourism Management, 25*(4), 471–481.

Turing, A. M. (1950). Computing machinery and intelligence. *Mind, 59*(236), 433–460.

van der Schalk, J., Hawk, S. T., Fischer, A. H., & Doosje, B. (2011). Moving faces, looking places: Validation of the Amsterdam Dynamic Facial Expression Set (ADFES). *Emotion, 11*(4), 907–920.

van Doorn, J., Mende, M., Noble, S. M., Hulland, J., Ostrom, A. L., Grewal, D., & Petersen, J. A. (2017). Domo Arigato Mr. Roboto: Emergence of automated social presence in organizational frontlines and customers' service experiences. *Journal of Service Research, 20*(1), 43–58.

Vassos, S., Malliaraki, E., Dal Falco, F., Di Maggio, J., Massimetti, M., Nocentini, M. G., & Testa, A. (2016). Art-Bots: Toward chat-based conversational experiences in museums. In F. N. a. A. S. Gordon (Ed.), *Proceedings of the 9th International Conference on Interactive Digital Storytelling, LNCS 10045* (pp. 433–437). Los Angeles: Springer.

Verhagen, T., Van Nes, J., Feldberg, F., & Van Dolen, W. (2014). Virtual customer service agents: Using social presence and personalization to shape online service encounters. *Journal of Computer-Mediated Communication, 19*(3), 529–545.

Vidal, L., Ares, G., & Jaeger, S. R. (2016). Use of emoticon and emoji in tweets for food-related emotional expression. *Food Quality and Preference, 49*(December), 119–128.

Viglione, S. (1970). Applications of pattern recognition technology. In J. M. Mendel & K. S. Fu (Eds.), *Adaptive, learning, and pattern recognition systems* (pp. 115–162). Cambridge, USA: Academic Press.

von der Pütten, A. M., Krämer, N. C., Gratch, J., & Kang, S. H. (2010). "It doesn't matter what you are!" Explaining social effects of agents and avatars. *Computers in Human Behavior, 26*(6), 1641–1650.

Walker, R. H., Craig-Lees, M., Hecker, R., & Francis, H. (2002). Technology-enabled service delivery. *International Journal of Service Industry Management, 13*(1), 91–106.

Walker, R. H., & Johnson, L. W. (2006). Why consumers use and do not use technology-enabled services. *Journal of Services Marketing, 20*(2), 125–135.

Wallis, P., & Norling, E. (2005). The trouble with chatbots: Social skills in a social world. *Virtual Social Agents, 29*, 29–36.

Walther, J. B., & D'Addario, K. P. (2001). The impacts of emoticons on message interpretation in computer-mediated communication. *Social Science Computer Review, 19*(3), 324–347.

Watson, D., Clark, L. A., & Tellegen, A. (1988). Development and validation of brief measures of positive and negative affect: the PANAS scales. *Journal of Personality and Social Psychology, 54*(6), 1063–1070.

Weizenbaum, J. (1966). ELIZA—a computer program for the study of natural language communication between man and machine. *Communications of the ACM, 9*(1), 36–45.

Westbrook, R. A., & Oliver, R. L. (1991). The dimensionality of consumption emotion patterns and consumer satisfaction. *Journal of Consumer Research, 18*(1), 84–91.

Wieseke, J., Geigenmüller, A., & Kraus, F. (2012). On the role of empathy in customer-employee interactions. *Journal of Service Research, 15*(3), 316–331.

Wirtz, J., Patterson, P. G., Kunz, W. H., Gruber, T., Lu, V. N., Paluch, S., & Martins, A. (2018). Brave new world: Service robots in the frontline. *Journal of Service Management, 29*(5), 907–931.

Wispé, L. (1986). The distinction between sympathy and empathy: To call forth a concept, a word is needed. *Journal of Personality and Social Psychology, 50*(2), 314–321.

Wispé, L. (1991). *The psychology of sympathy*. New York, USA: Plenum Press.

Wolf, K. (2015). Measuring facial expression of emotion. *Dialogues in Clinical Neuroscience, 17*(4), 457–462.

Wood, A. M., Linley, P. A., Maltby, J., Baliousis, M., & Joseph, S. (2008). The authentic personality: A theoretical and empirical conceptualization and the development of the Authenticity Scale. *Journal of Counseling Psychology, 55*(3), 385–399.

Wood, M. (2005). Bootstrapped confidence intervals as an approach to statistical inference. *Organizational Research Methods, 8*(4), 454–470.

Wünderlich, N. V., & Paluch, S. (2017). A nice and friendly chat with a bot: User perceptions of AI-based service agents. In *Proceedings of the 38th International Conference on Information Systems (ICIS 2017)*. Seoul, South Korea.

Yan, R., Song, Y., & Wu, H. (2016). Learning to respond with deep neural networks for retrieval-based human-computer conversation system. In *Proceedings of the 39th International ACM SIGIR Conference on Research and Development in Information Retrieval* (pp. 55–64). Pisa, Italy: Association for Computing Machinery.

Yoo, J., & Arnold, T. J. (2016). Frontline employee customer-oriented attitude in the Presence of job demands and resources. *Journal of Service Research, 19*(1), 102–117.

Yoon, S., Lee, S. S., Lee, J. M., & Lee, K. (2014). Understanding notification stress of smartphone messenger app. In *CHI'14 Extended Abstracts on Human Factors in Computing Systems* (pp. 1735–1740).

Zeithaml, V. A., Berry, L. L., & Parasuraman, A. (1996). The behavioral consequences of service quality. *Journal of Marketing, 60*(2), 31–46.

Zhang, Y., Zhang, J., & Sakulsinlapakorn, K. (2020). Love becomes hate? Or love is blind? Moderating effects of brand love upon consumers' retaliation towards brand failure. *Journal of Product & Brand Management, 30*(3), 415–432.